Al-Ashab Collage Artwork by Rana Ibrahim

A HERBAL of
IRAQ
أعشابٌ من العراق

A HERBAL of IRAQ

أعشابٌ من العراق

A herbal inspired by the work of Abdul Jaleel Ibrahim Al-Quragheely, the Doctor of Al-Ashab

عشبة مستوحاة من عمل عبد الجليل إبراهيم القره غولي الدكتور العشاب

Edited by Shahina A. Ghazanfar and Chris J. Thorogood
Arabic editor Rana Ibrahim

Kew Publishing
Royal Botanic Gardens, Kew

First published in 2022 by
Royal Botanic Gardens, Kew,
Richmond, Surrey, TW9 3AB, UK
www.kew.org

ISBN 978-1-84246-763-3

Distributed on behalf of the Royal Botanic Gardens, Kew in North America
by the University of Chicago Press, 1427 East 60th St, Chicago, IL 60637, USA

British Library Cataloguing in Publication Data

A catalogue record for this book is available from the British Library

Design and page layout: Christine Beard
Project manager: Lydia White
Production manager: Jo Pillai
Copy-editing: Matthew Seal
Proofreading: Sharon Whitehead

Cover illustration: *Glossostemon bruguieri* © C. J. Thorogood
Endpapers: *Al-Ashab Collage Artwork* by Rana Ibrahim. www.iwaw19.com © Rana Ibrahim

Printed in Great Britain by Bell & Bain Ltd, Glasgow

For information or to purchase all Kew titles please visit
shop.kew.org/kewbooksonline or email publishing@kew.org

Kew's mission is to understand and protect plants and fungi, for the wellbeing of people and the future of all life on Earth.

Kew receives approximately one third of its funding from Government through the Department for Environment, Food and Rural Affairs (Defra). All other funding needed to support Kew's vital work comes from members, foundations, donors and commercial activities, including book sales.

Rana Ibrahim's dedication

This book is dedicated to my dear mother, Nazik Saeed Al-Jarrah, faithful wife to my martyr father, Abdul Jaleel Ibrahim Al-Quragheely, and to all my sisters, brothers, and other members of my honourable family.

إهداء رنا إبراهيم: أهدي هذا الكتاب إلى أمي العزيزة، نازك سعيد الجرّاح، الزوجة الوفية لأبي الشهيد، عبد الجليل إبراهيم القره غولي، ولشقيقاتي وأشقائي وكل فردٍ من أفراد أسرتي الكريمة.

CONTENTS

المحتويات

Anethum graveolens

Bryonia multiflora

ACKNOWLEDGEMENTS

We thank Nazik Saeed Al-Jarrah for writing Abdul Jaleel Ibrahim Al-Quragheely's story, which has been translated by Yasmin Mahmud, grandchild of Abdul Jaleel Ibrahim Al-Quragheely. We thank Stephen Harris for his generous contribution of supporting imagery from the University of Oxford Herbaria, and the Illustrations Team, Library and Archives, Royal Botanic Gardens, Kew for supplying several images for this Herbal. Our sincere thanks also to Emma Webster and Harriet Warburton for their support in the delivery of the project. Most of all we thank Rana Ibrahim, the faithful custodian of the life work of Abdul Jaleel Ibrahim Al-Quragheely, without whom this herbal flora would never have been possible. This work was generously supported by the John Fell Fund, an internal research fund financed by Oxford University Press.

شُكر و تقدير

نتقدم بالشكر إلى نازك سعيد الجرَّاح على كتابة قصة عبد الجليل إبراهيم القره غولي، وترجمت القصة ياسمين محمود حفيدة عبد الجليل إبراهيم القره غولي .كما نشكر ستيفن هاريس لمساهمته الكريمة بالصور الشارحة من جامعة أكسفورد هيرباريا (مجموعة النباتات المجففة المملوكة لجامعة أكسفورد)، كما نتوجّه بخالص الشكر لإما ويبستر وهارييت واربورتون لتقديم الدعم اللازم لتسليم هذا المشروع، ونُقدِّم أسمى آيات الشكر والعرفان لرنا إبراهيم الراعية الأمينة لأهم أعمال عبد الجليل إبراهيم القره غولي، والتي لولاها ما خرجت هذه النباتات العشبية للنور. لقد لاقى هذا العمل دعمًا عظيمًا من صندوق جون فيل، وهو صندوق لدعم الأبحاث الداخلية مُموَّلٌ من مطبوعات جامعة أكسفورد.

Prunus armeniaca (apricot)

FOREWORD

by Abdul Jaleel Ibrahim Al-Quragheely, the Doctor of Al-Ashab†

All living organisms exist on our planet according to the laws of nature – laws that apply equally to man and all other creatures, including those we may be yet to encounter. The plant and animal kingdoms live side by side – their lives are intertwined. Let us consider the apricot tree, its delicious fruit and at its core, the seed – a plant within a plant. We may ask ourselves how the fruit developed: did the tree make the seed and its hard shell? Or was it, in part at least, the farmer who helped produce the fruit? We may conclude that neither the tree nor the farmer could create the characteristics of the apricot fruit in the form that we appreciate today – a form that has been nurtured by people for thousands of years.

Herbal medicine is as old as humankind itself. The science of herbal medicine is still evolving and there is much to discover still. The herbal physician must work hard to acquaint himself with a multitude of medicinal plants and to identify the complex characteristics of human disease that they may be effective remedies for. The methods of using medicinal plants and herbs are vast and include application in the form of extracts, solutions, powders and oils, as well as mixtures of all of these. Herbalists are able to convert plants from their natural states into powerful and effective remedies – concentrating them and enhancing them while paying close attention to the processes of their preparation, drying, packaging and storage. The responsibilities of the herbalist are equally vast: they may range from identifying plant-based salves for the treatment of acne to tonics for soothing rheumatic pain, alleviating asthma, extracting kidney stones and treating diabetes.

My work has sought to raise awareness of the work of Arabic scholars on medicinal herbs, inspired by prophetic medicine and Islamic medicine, and their importance to humankind. During the course of preparing this herbal flora I set out to confirm the identity of medicinal plants that are still used in Iraq today, of which there are many. I sought to understand the importance of the vital active substances within the plants, and to determine which of them have the power to bring about effective treatment. Over the years, my work has explored the principles of preparing medicinal herbal extracts from several plants in combination. It must be stressed that the efficacy of a given herbal concoction relies heavily on the practical

experience of the herbalist and his level of experience in the composition, preparation and administration of the medicine.

Iraq is blessed with a richness of medicinal plants of all kinds. Its varied environmental conditions enable the cultivation of many plant species that can be used in medicine. Its plants must be valued and harvested judiciously and sustainably. Let us consider the moghat plant ('erok orab kuzzi', *Glossostemon bruguieri*), a shrub with dark-coloured roots, and a plant of significant nutritional and traditional medicinal value. The Hamrin Mountains are now all but devoid of this much-cherished medicinal herb. Not only must we value these important plants, but we must appreciate that herbal medicine, both in Iraq and elsewhere, is multifaceted and spans many disciplines: agriculture – the cultivation of plants in the field; research – understanding the properties of plants in the laboratory and harnessing the knowledge obtained in herbal medical institutes, colleges and universities; and finally, experience – in the form of the vital work carried out by herbalists in their communities. Plants have untapped potential to cure and change people's lives. We must value them, research them and conserve them for the benefit of all humankind.

Ricinus communis

مُقدمة بقلم الدكتور العشَّاب عبد الجليل إبراهيم القره غولي

يخضع وجود الكائنات الحيّة كافة على كوكبنا لقوانين الطبيعة؛ وهي قوانين تنطبق بالتساوي على الإنسان والمخلوقات الأخرى كافة، بما في ذلك تلك المخلوقات التي ربما لم نتعرّف عليها بعد. تعيش الممالك النباتية والحيوانية جنبًا إلى جنب؛ أي أن حياتهما متشابكة؛ فإذا ما نظرنا إلى شجرة المشمش، فإنها ذات ثمرة لذيذة وداخلها البذرة - نباتٌ داخل نبات. وقد يراودنا سؤال حول كيفية نمو الثمرة: هل أثمرت الشجرة البذرة وقشرتها الصلبة؟ أم أن المزارع هو الذي ساعد، ولو جزئيًا على الأقل، في إخراج الفاكهة؟ وقد نصل إلى استنتاجٍ مفاده أنه ليس بوسع الشجرة ولا المزارع خلق خصائص ثمار المشمش بالشكل الذي نراه اليوم - وهو الشكل الذي تعهده البشر بالرعاية لآلاف السنين.

إن طب الأعشاب قديم قدم الجنس البشري نفسه، وما زال علم طب الأعشاب يتطوّر وما زال هناك الكثير لاكتشافه. وينبغي لطبيب الأعشاب أن يسعى حثيثًا للتعرّف على الكثير من النباتات الطبية وتحديد الخصائص المعقدة للأمراض التي تصيب الإنسان والتي قد تكون هذه النباتات علاجات فعَّالة لها. ثمة طرق عدة لاستخدام النباتات الطبية والأعشاب تشمل استخدام هذه النباتات والأعشاب في شكل مستخلصات ومحاليل ومساحيق وزيوت وخلائط من كل ذلك؛ حيث إنه بمقدور المُعالجين بالأعشاب تحويل النباتات من حالتها الطبيعية إلى علاجات قوية وفعَّالة - بتركيزها وتعزيزها؛ مع إيلاء عناية جادة لعمليات تحضيرها وتجفيفها وتعبئتها وتخزينها. وتقع على عاتق المُعالج بالأعشاب مسؤوليات بهذه الضخامة بدءًا من تحديد المستحضرات القائم تركيبها على نباتات ما بين مرهم لعلاج حب الشباب إلى مقويات لتسكين الآلام الروماتيزمية، وتخفيف حالات الربو، واستخراج حصوات الكلى، وعلاج داء السكري.

يهدف عملي إلى زيادة الوعي بجهود العلماء العرب في مجال الأعشاب الطبية والمستوحاة من الطب النبوي والطب الإسلامي وأهمية هذه الجهود للبشرية. وقد شرعت أثناء إعداد هذا الكتاب المعني بالنباتات العشبية في التأكد من هوية النباتات الطبية التي ما زالت مستخدمة في العراق إلى يومنا هذا، وهي كثيرة، كما سعيت إلى فهم أهمية المواد الفعَّالة الحيوية في النباتات، ومعرفة أيها له القدرة على العلاج الفعَّال. لقد دأبت على مر السنين على استكشاف مبادئ تحضير المستخلصات العشبية الطبية من عدة نباتات مخلوطة، ولا بدّ من التأكيد على أن فعالية خليط عشبي مُعين تعتمد اعتمادًا بالغًا على الخبرة العملية للمُعالج بالأعشاب ومستوى خبرته في تركيب هذا الخليط وتحضيره ووصفه للمريض.

حبا الله العراق بنباتات طبية من شتى الأنواع؛ فظروفها البيئية المتنوّعة مكّنتها من زراعة الكثير من فصائل النباتات التي يمكن استخدامها كدواءٍ، ومن ثمَّ لا بدّ من أن تحظى نباتات العراق بالاهتمام والعناية بحصادها واستدامته. دعونا نتناول نبات المُغات (عرب قوزي، أو *Glossostemon bruguieri*)؛ حيث ينمو نبات

Glossostemon bruguieri

المُغات في شكل شجيرة ذات جذور داكنة اللون وله قيمة غذائية وطبية منذ القدم. ولقد أصبحت تلال حمرين الآن خالية تمامًا من هذه العشبة الطبية ذات القيمة العالية. ليس علينا أن نُقدِّر هذه النباتات المُهمة فحسب؛ بل يجب أن نُقدِّر أن طب الأعشاب في العراق وأماكن أخرى متعدد الجوانب وممتد إلى كثيرٍ من التخصصات: الزراعة - زراعة النباتات في الحقل، والبحث - فهم خصائص النباتات في المختبر وتسخير معارف المعاهد والكليات والجامعات الطبية العشبية لخدمة طب الأعشاب، وأخيرًا، الخبرة – المتمثلة في العمل الحيوي الذي يؤديه المُعالجون بالأعشاب في مجتمعاتهم. وتحظى النباتات بقدراتٍ غير مستغلة في مداواة حياة البشر وتغييرها، وعلينا أن نُثمِّن هذه القدرات ونُوجِّه الجهود البحثية لها ونحافظ عليها لصالح البشرية جمعاء.

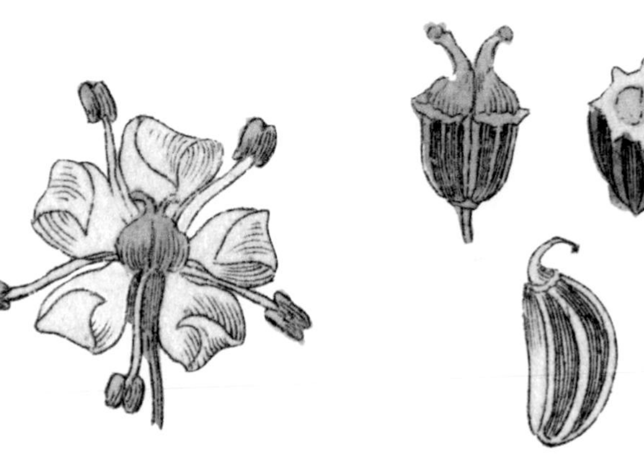

Apium graveolens

DUPLICATE

FLORA OF KUWAIT

Family Chenopodiaceae No. 11715

Species Hammada salicornica (Moq.) Iljin

County Sobiyah Road, on the way to Sobiyah,

Locality 10km. from T.V. camp, left side of road.

47°50E 29° 31 N Altitude

Collectors A. Rawi & H. El-Kholy Date 18 Nov. 1982

Notes red flowers

Haloxylon salicornicum, a shrub that is characteristic of the deserts of Iraq.

An oak (*Quercus* sp.) collected in the mountains of Iraq in 1859.

INTRODUCTION

THE FLORA OF IRAQ

Biogeography

Iraq lies in the centre of the Middle East, a transcontinental region that is home to some of the earliest human civilizations to domesticate food crops. The region is defined by extensive rocky and sandy deserts, as well as highland plateaus and mountain ranges. While rainfall and temperature vary considerably across the region, the climate is predominantly hot, dry and arid, and prone to dust and sand storms. Temperatures in summer are typically around 30°C (86°F) but have reached 50°C (122°F) in the Iraqi city of Basra, the highest temperature recorded across major Middle Eastern cities. Even winter temperatures can exceed 25°C (77°F) in lowland desert areas of Iraq. Two major rivers flow through this hot, dry landscape: the Euphrates and the Tigris. The Euphrates is the longest river in the Middle East and features prominently in ancient history. Its source lies in the Caucasus Mountains of Armenia, from where it flows across east-central Turkey, through Syria and Iraq, where it joins the Tigris, and into the Persian Gulf.

Deserts cover much of Iraq and the surrounding Middle East. The Arabian Desert (2,330,000 km^2 or 890,000 sq. miles) stretches from Yemen to the Persian Gulf, and from Oman to Jordan and Iraq, and is one of the world's largest deserts. It encompasses vast stretches of hyper-arid dunes that belong to an area called Ar Rub' Al Khali, or 'The Empty Quarter'. Also part of the Arabian Desert is the Syrian Desert (518,000 km^2 or 200,000 sq. miles), which dissects Jordan, Syria, Saudi Arabia and western Iraq. This is a region of extremely low rainfall that is inhabited by nomads who raise cattle and camels. Mountains in Iraq, as elsewhere in the Middle East, are a refuge for plant diversity in an otherwise barren landscape. The Zagros mountain range, for example, is an area of species richness for Iran, Iraq and Turkey. The range begins in northwest Iran, runs the length of the west and southwest Iranian plateau, and ends at the Strait of Hormouz. Precious remnants of formerly dominant oak (*Quercus brantii*) forest occur here, along with an extraordinary assemblage of ancestral food plants including almond, apricot, barley, grape, lentil, pistachio, pomegranate, walnut and wheat (Frey *et al.*, 1986).

The Middle East lies at the crossroads of Asia, Africa and Europe. The region comprises a transition zone of two major biogeographic 'ecozones' called the Palaearctic and Afrotropical (Olson *et al.*, 2001). These ecozones are separated by the Arabian Desert, where unique transitional floras can be found. The confluence of these distinct biogeographic units has resulted in a high biodiversity in the region, but this is also under severe threat, with only an estimated 4% of the original vegetation remaining. For this reason, the Middle East is designated one of the world's 36 Biodiversity Hotspots (Olson *et al.*, 2001; Mildrexler *et al.*, 2005).

Flora

The flora of the Middle East is rich and diverse, owing to its complex and varied geography and climate. Although difficult to quantify because the floristics are incomplete or even embryonic for parts of the region, an estimate of 12,500 species is plausible (Ghazanfar & McDaniel, 2016). The flora of Iraq is particularly rich and diverse, owing to the country's diverse habitats of deserts, plains and mountains. Floristic work is incomplete in Iraq and so, as is the case elsewhere in the Middle East, the number of plant species that occur there is difficult to ascertain with certainty. Efforts to document the *Flora of Iraq* started in 1965 as a collaborative effort between the Royal Botanic Gardens, Kew and the Ministry of Agriculture, Baghdad. Progress on this seminal *Flora* was interrupted by political change in Iraq in recent decades, but work has recommenced. On the basis of the work carried out to date, Ghazanfar and McDaniel (2016) produced the first quantitative analysis of the flora of Iraq, estimating that some 3,300 flowering plants species from 908 genera belonging to 136 families occur in the region. We now have a good estimate of about 3,950 taxa.

The plant families with the greatest representation are the daisy family (Asteraceae) with over 400 species, and the pea family (Fabaceae) with 393 species, followed by the grass family (Poaceae: 264 species), cabbage family (Brassicaceae: 195 species) and carrot family (Apiaceae: 155 species); a similar hierarchy is prevalent across the Middle East and Mediterranean Basin. Annuals make up more than a third of the flora. The number of endemic plants (those found only in the region) is possibly low in Iraq, at an estimated 5–6% of the flora (181 species), most of which are found in the country's northern mountains. Indeed, the northern mountainous district contains 40% of the country's flora because of its climate, isolation from land use change, and location as a point of confluence of the Turkish and Iranian floras. The Zagrosian foothills and mountains, which range between 1,000 m (3,300 ft) and 1,700 m (5,500 ft), are particularly rich in endemics relative to other districts in Iraq (Youssef, 2020). Iraq's Central Alluvial Plains also host a rich assortment of halophytes (salt-tolerant plants) and annuals.

Plant communities

Iraq is bordered by mountainous parts of Turkey and Iran to the north and east, respectively, by the Saudi Arabian desert to the south and by Syria in the west, where the land is called Albadia and is characterized by low rainfall (Al-Douri, 2014). The flora of Iraq is linked to those of each of these countries and can be divided loosely into four major vegetation zones, each with their own distinct plant communities: desert, mixed shrubland, mountain forest and alpine forest (Guest & Al-Rawi, 1966; Ghazanfar & McDaniel, 2016).

Desert (350,000–400,000 km² or 135,000–154,000 sq. miles) comprises the Southern Desert and Sub-desert (steppe) that borders Saudi Arabia, Jordan, Syria and Kuwait. This is by far the largest zone in Iraq by surface area. The climatically defined Western Desert District, Southern Desert District, Lower Jazira District and Alluvial Plains District of Iraq all fall within this region. Here, vast plains of sand and gravel are dissected in places by depressions, rocky outcrops, river systems and marshes in which most of the area's plants occur. The region also encompasses the sabkha (flatland) and saline areas of the alluvial plains of Lower Mesopotamia, and cultivated land. Annual rainfall across the region is low, typically in the range of 75–150 mm (3–6 in), and the level of salinity is high. Consequently the vegetation mainly comprises sparse, open desert scrub and halophytic (salt-tolerant) plant communities. Black saxaul (*Haloxylon ammodendron*), a shrub planted globally in areas prone to desertification, dots the sandy deserts and unstable dunes, along with other gnarled xerophytic (drought-tolerant) shrubs such as *Calligonum comosum* and clumping desert bunchgrass (*Panicum turgidum*). Another xerophytic shrub, *Haloxylon salicornicum*, is characteristic of the region and is also widespread across the deserts of the Arabian Peninsula. Succulent subshrubs that frequent the sandy soil overlying saline flats include *Zygophyllum coccineum* and *Anabasis articulata*. Another desert shrub that has value as a grazing plant is *Halocnemum strobilaceum*, which dominates the sabkha areas prevalent on saline mud-flats in southwest Iraq. In some areas, the desert shrubs listed above are parasitized by the imposing yellow and purple leafless spikes of 'desert hyacinth' (*Cistanche tubulosa*), which makes a desultory appearance after late winter rainfall.

Pistacia khinjuk collected in a cemetery in Iraq in 1933.

Sub-desert scrub is a habitat for numerous annuals that spring up after winter rain, including various desert milkvetches (*Astragalus* species such as *A. tribuloides*), grasses such as *Stipa capensis* and the daisy relative – a form of 'resurrection plant' – *Pallenis hierochuntica*. Sand button or 'als'dan' (*Neurada procumbens*), a prostrate annual widespread across Middle Eastern dunes and deserts, produces distinctive spiny fruits; elsewhere the plant has been used as a camel food and in traditional Arabic medicine to treat diarrhoea and dysentery (Chelalba *et al.*, 2020).

The Central Alluvial Plains and Southern Marsh District of Iraq comprise a complex of shallow freshwater lakes, swamps, marshes and seasonally flooded plains between the Tigris and Euphrates rivers. These wet plains are an important wintering site for migratory birds in Eurasia and an oasis for biodiversity. Over 800 species of plant are estimated to occur here, a handful of which grow nowhere else.

Mixed shrubland (65,000 km^2 or 25,100 sq. miles) includes the dry and moist steppes in northern Iraq, bordering Syria to the east and Iran to the west. In Iraq, it constitutes the upper plains to the foothills of the mountains, gravel plains, rocky outcrops and smallholdings farmed for winter crops. The climatically defined Upper Jazira District, Kirkuk District, Arbil District, Nineveh District and Persian Foothills District fall within this region. Annual rainfall is far higher in the mixed shrubland region than in the desert region, in the ranges of 200–350 mm (8–14 in) and 350–500 mm (14–20 in). Its steppes are characterized by sparse, short grassland with small, scattered shrubs. Characteristic vegetation includes meadow grasses such as *Poa bulbosa* and *Carex stenophylla* along with colourful splashes of Persian buttercup (*Ranunculus asiaticus*) and crown anemone (*Anemone coronaria*) in the spring. Remnant plant communities of *Artemisia herba-alba* and *Achillea conferta* occur in the steppe region, and rocky slopes are home to bulbs including *Ornithogalum neurostegium*, *Gladiolus italicus*, *Bellevalia* species and *Leopoldia longipes*. Dry steppe open grasslands are dotted with small trees such as *Pistacia* here and there, although the natural vegetation has been replaced extensively by grazed and agricultural land.

Mountain forest (30,000 km^2 or 11,500 sq. miles) includes the northern mountains that border Syria, Turkey and Iran, including the higher-altitude areas of Jabal Sinjar, and falls within the important Zagros Mountains Forest Steppe ecoregion. Annual precipitation is in the region of 700–1,400 mm (28–55 in), partly in the form of snow cover. The region constitutes open to closed forest that transitions into a 'thorn cushion-type' vegetation. Forests are dominated by various species of oak (*Quercus aegilops*, *Q. brantii*, *Q. infectoria*, *Q. libani*) interspersed with Persian turpentine tree (*Pistacia atlantica*) and related *P. khinjuk*, as well as stands of pine forest (*Pinus brutia* var. *eldarica*). The thorn-cushion plant community comprises an open shrubland dominated by large, spiny domes of milkvetch (*Astragalus* spp.), associated with *Daphne acuminata*, *Lonicera arborea* and various low thorny shrubs such as *Acantholimon*, *Acanthophyllum* and *Cousinia*. Much of the original forest has been cleared, and closed-canopy forests occur now in just a few inaccessible locations.

Alpine forest (100 km^2 or 39 sq. miles) includes the alpine (2,750–3,730 m or 9,000–12,200 ft) zone of the northern mountains of the country, encompassing the Amadiya District, Rowanduz District and Sulaimaniya District. Annual precipitation is over 1,000 mm (39 in) in these districts, largely in the form of snow. As in the previous zone, alpine forest falls within the important Zagros Mountains Forest Steppe ecoregion but its higher-altitude slopes host a different form of vegetation dominated by low perennial herbs and shrubs of the families Asteraceae, Brassicaceae, Fabaceae, Lamiaceae, Plumbaginaceae and Poaceae. Although small in area compared with the previous zones, the alpine forest region is home to a disproportionately high number of rare and endemic plant species. The northern mountains of Iraq in Kurdistan are an extension of the alpine system that runs eastwards from the Balkans through southern Turkey, northern Iraq, Iran and Afghanistan, eventually reaching the Himalayas. In Iraq, the mountains in the Amadiyah, Rowanduz and Sulimaniyah districts have the highest number of endemic plant species, in particular Jebel Avroman, a mountain range (2,500 m or 8,200 ft) running along the Iranian frontier 15–25 km (9 to 15 miles) northeast and north of Halabja and Pira Magrun, and also a high mountain peak (2,800 m or 9,200 ft) about 30 km (19 miles) northwest of Sulaimaniyah.

TRADITIONAL HERBAL MEDICINE IN IRAQ

Plant-based remedies date back to the beginning of mankind. Traditional medicine in Iraq has a particularly long and rich history, which can be traced back to the Sumerian (4000–1970 BCE) and Babylonian and Assyrian periods (1970–549 BCE). To this day, both urban and rural communities depend on this rich heritage of knowledge across all districts, including the desert, steppe and uplands of Iraq (Al-Douri, 2014). A similar situation exists beyond Iraq: according to the World Health Organization (WHO), 80% of populations in developing countries meet their basic medical needs with traditional herbal medicine. Although the availability of modern medicine has increased in Iraq, traditional herbal medicine is a vital part of the healthcare system, especially in desert areas, where it is practised by 'attars' (herbalists) in the country. Despite the importance of herbalists in the community, their work is often undocumented, and the rich heritage of traditional herbal medicine is at severe risk of being lost.

Research into the practising of herbal medicine by traditional healers in the Sulaymaniyah and Erbil Governorates of Kurdistan shows that numerous plant species from across at least 34 families are harvested in this region. Leaves, and to a lesser extent, flowers and seeds, were harvested most frequently from the mint family (Lamiaceae), carrot family (Apiaceae), daisy family (Asteraceae) and pea and bean family (Fabaceae) (Ahmed, 2016; Naqishbandi, 2014). The most common preparation method here was found to be decocting (the concentration or extraction of a plant by boiling in a liquid), although plants were also consumed as a vegetable or ingested in powder form. In Kurdistan, plant-based remedies are the only form of medicine in remote mountain villages, where knowledge of the effects and uses of herbs is not only the property of herbalists, as in other traditional medicinal systems, but rather it is integral to the cultural heritage of families and is passed down the generations. Any market, 'souk' and bazaar contains a section where medicinal wild plants are sold, for example, fresh wild tulip bulbs can be bought as a traditional painkiller (Amin *et al.*, 2016). Herbalists' shops in the region trade wild-sourced and cultivated plant-based remedies from all over the Middle East and Asia Minor. While most of the plants traded have been used since antiquity, herbalists in 2011 were reporting a steady increase in trade in conditions of greater economic stability and up to 64% of medicinal plants were imported from outside Iraq at that time (Mati & de Boer, 2011). It should be noted that plants in Iraq also have various other traditional uses besides herbal medicine, including uses as foods, tools, gums, fodder, tanning and dyes (Ahmad & Askari, 2015).

Peganum harmala

مُقدمة

نباتات العراق

الجغرافيا الأحيائي

يقع العراق في قلب الشرق الأوسط، وهي منطقة تقاطع القارات التي كانت مهدًا لبعض أقدم الحضارات الإنسانية على مستوى العالم التي نشأت على المحاصيل الغذائية. وتشتهر هذه المنطقة بالصحارى الصخرية والرملية الشاسعة والهضاب والمرتفعات والسلاسل الجبلية، بينما يتنوّع سقوط الأمطار ودرجات الحرارة تنوعًا ملحوظًا على مستوى المنطقة، ويسودها مناخ حار جاف قاحل، والمنطقة عرضة للعواصف الترابية والرملية. وعادة ما تصل درجات الحرارة في الصيف إلى نحو 30 درجة مئوية، ولكنها تبلغ 50 درجة مئوية في مدينة البصرة العراقية، وهي أعلى درجة حرارة سُجلت على مستوى المدن الرئيسية في الشرق الأوسط. وقد تتجاوز درجات الحرارة في الشتاء أيضًا 25 درجة مئوية في المناطق الصحراوية المنخفضة في العراق. يشق هذه الطبيعة الحارة القاحلة نهران رئيسيان؛ الفرات ودجلة؛ حيث يُعد الفرات أطول أنهار الشرق الأوسط ويبرز ذكره في التاريخ القديم، وينبع نهر الفرات من جبال القوقاز في أرمينيا، ويتدفق عبر شرق وسط تركيا، مارًا بسوريا والعراق، حيث يلتقي بنهر دجلة، ويمتد نحو الخليج الفارسي.

تغطي الصحارى معظم أنحاء العراق ومناطق الشرق الأوسط المُحيطة، وتمتد الصحراء العربية (٣٣٢٠٠٠٠ كم٢) من اليمن إلى الخليج الفارسي، ومن سلطنة عُمان إلى الأردن والعراق، وتعد إحدى أكبر الصحارى في العالم، وهي تغطي مساحات شاسعة من الكثبان الرملية شديدة الجفاف والتي تنتمي إلى منطقة يطلق عليها «صحراء الربع الخالي»، كما تعد الصحراء السورية (518,000 كم2) والتي تتقاطع مع الأردن وسوريا والمملكة العربية السعودية وغرب العراق جزءًا من الصحراء العربية. يقل هطول الأمطار بهذه المنطقة التي يسكنها البدو الذين يربون الماشية والإبل. وتعد جبال العراق، شأنها في ذلك شأن أرجاء الشرق الأوسط كافة، مواطن للتنوّع النباتي وسط طبيعة قاحلة؛ فعلى سبيل المثال، سلسلة جبال زاجروس منطقة غنية بأنواع النباتات في إيران والعراق وتركيا، ويبدأ نطاقها من شمال غرب إيران ويمتد بطول الهضبة الإيرانية الغربية والجنوبية الغربية وينتهي عند مضيق هرمز. توجد في هذه المنطقة بقايا ثمينة من غابات البلوط (*Quercus brantii*) التي امتدت على هذه الأراضي فيما سبق، جنبًا إلى جنب مع مجموعة هائلة من النباتات الغذائية التي نبتت فيها منذ القدم؛ بما في ذلك اللوز والمشمش والشعير والعنب والعدس والفستق والرمان والجوز والقمح (فراي وآخرون، 1986).

يقع الشرق الأوسط على مفترق الطرق بين آسيا وأفريقيا وأوروبا، ويضم منطقة انتقالية تتألف من «منطقتين بيئيتين» رئيسيتين من المناطق الجغرافية الحيوية، هما المنطقة القطبية الشمالية القديمة «Palaearctic» والإقليم المداري الإفريقي «Afrotropical» (أولسون وآخرون، 2001)؛ حيث تفصل

الصحراء العربية هاتين المنطقتين البيئيتين ويمكن العثور على نباتات انتقالية فريدة من نوعها. وقد أدى التقاء هذه الوحدات الجغرافية الحيوية المميزة إلى زيادة التنوّع البيولوجي في المنطقة التي تتعرض أيضًا لتهديد شديد، مع بقاء ما يقدر بنحو ٤٪ فقط من الحياة النباتية الأصلية بها؛ ولهذا السبب تُصنَّف منطقة الشرق الأوسط ضمن 36 منطقة ذات تنوّعٍ بيولوجيٍ شديدٍ حول العالم (أولسون وآخرون، 2001؛ ميلدركسلر وآخرون، 2005).

النباتات

تمتاز نباتات منطقة الشرق الأوسط بالثراء والتنوّع بفضل جغرافيا المكان ومناخه المعقد والمتنوّع، وعلى الرغم من صعوبة تحديد كمياتها لأن دراسات الجغرافيا النباتية المعنية بأنواع النباتات الموجودة بالمنطقة غير مكتملة أو بدائية في بعض أنحاء المنطقة، فقد رصد تقديرٌ مقبولٌ وجود 12,500 نوعٍ من أنواع النباتات في المنطقة (غضنفر وماك دانيال، 2016). وتمتاز النباتات في العراق على وجه الخصوص بالثراء والتنوّع بفضل تنوّع البيئات بالمنطقة ما بين صحارى وسهول وجبال. جديرٌ بالذكر أن دراسات الجغرافيا النباتية في العراق غير مستوفاة؛ ولذلك من الصعب تحديد عدد أنواع النباتات الموجودة في العراق على وجه اليقين، شأنها في ذلك شأن باقي أنحاء الشرق الأوسط. وقد دُشنت الجهود المشتركة في عام 1965 بين حدائق النباتات الملكية (كيو) ووزارة الزراعة في بغداد بهدف توثيق نباتات العراق، ولكن التغيّر السياسي الذي طرأ على العراق في العقود الأخيرة اعترض تقدم هذه الجهود الفاعلة في هذا المجال، بيْد أنها قد أُستؤنفت مُجددًا. وبناءً على العمل الذي تم حتى الآن، أجرى غضنفر وماك دانيال (2016) أول تحليلٍ كميٍّ لنباتات العراق؛ حيث قدَّرا أن هناك نحو 3,300 نوعٍ من النباتات المزهرة تمثل 908 جنسٍ تنتمي إلى 136 فصيلة موجودة في المنطقة. ولدينا الآن تقدير جيد لنحو 3,950 مجموعة مُصنَّفة. تُعد الفصيلة النباتية الأعلى تمثيلًا الفصيلة النجمية (Asteraceae) وتضم ما يزيد عن 400 نوعٍ، والفصيلة البقولية (Fabaceae) وتضم 393 نوعًا، تليها الفصيلة النجيلية (Poaceae: تضم 264 نوعًا)، ثم الفصيلة الكرنبية (Brassicaceae: تضم 195 نوعًا) والفصيلة الخيمية (Apiaceae: تضم 155 نوعًا)؛ ويسود منطقة الشرق الأوسط وحوض البحر المتوسط تسلسلٌ مماثلٌ؛ حيث تُمثِّل النباتات الحولية أكثر من ثلث النباتات. ومن المحتمل أن ينخفض عدد النباتات المتوطنة (الموجودة في المنطقة فقط) في العراق لنحو 5–6% (181 نوعًا)، ومعظمها في الجبال الموجودة بشمال البلاد. وفي واقع الأمر، تضم منطقة الجبال بشمال العراق 40% من النباتات الموجودة في العراق. ويرجع ذلك إلى مناخ المنطقة، وعزلتها عن التغيّر الناجم من استخدام الأراضي، وكونها نقطة التقاء للنباتات التركية والإيرانية. وتتراوح سفوح جبال زاجروس وقممها بين 1000 و1700 متر، وتمتاز بثرائها بالنباتات المتوطنة على وجه الخصوص إذا ما قورنت بغيرها من مناطق العراق (يوسف، 2020)، كما تضم منطقة السهول الرسوبية الوسطى في العراق مجموعة متنوّعة غنية من النباتات الملحية (نباتات تتحمل الملح) والنباتات الحولية.

التجمعات النباتية

يحدّ العراق من جهة الشمال والشرق الأجزاء الجبلية من تركيا وإيران على التوالي، ومن الجهة الجنوبية الصحراء العربية السعودية، ومن الغرب سوريا؛ حيث تُسمى هذه الأرض «البادية» وتتسم بقلة هطول الأمطار (الدوري، 2014). وترتبط نباتات العراق بالنباتات الموجودة في كل بلدٍ من هذه البلدان ويمكن تقسيمها بشكل غير محددٍ إلى أربع مناطق نباتية رئيسية، لكل منها مجتمعاتها النباتية المتميزة: الصحراء، والسهوب، والغابات الجبلية، وجبال الألب (جيست والراوي، 1966؛ غضنفر وماك دانيال، 2016).

الصحراء (40,000–350,000 كم²) – تتكوّن من الصحراء الجنوبية وشبه الصحراء (السهوب) التي تحدّ المملكة العربية السعودية والأردن وسوريا والكويت وتعد أكبر منطقة في العراق من حيث مساحة السطح. تقع الصحراء الغربية ذات الطبيعة المناخية المُحدَّدة ومنطقة الصحراء الجنوبية والجزيرة السفلى ومنطقة السهول الرسوبية في العراق ضمن هذه المنطقة. وتمتد سهول شاسعة المساحة من الرمال والحصى بفعل المنخفضات والنتوءات الصخرية ومنظومات التصريف والأهوار، حيث تنمو معظم نباتات المنطقة.
وتضم هذه المنطقة أيضًا مناطق السبخة والمناطق المالحة الموجودة في السهول الرسوبية في أسفل بلاد ما بين النهرين، كما تضم الأراضي المزروعة. وينخفض المعدل السنوي لهطول الأمطار على جميع أنحاء المنطقة، وعادة ما يتراوح بين 150–75 مم، ويرتفع مستوى الملوحة؛ ومن ثمَّ فإن الغطاء النباتي يتألف بشكل أساسي من شجيرات صحراوية وتجمعات نباتية ملحية (تتحمل الملوحة) متفرقة ومتناثرة؛ على سبيل المثال نبات الساكسول الأسود (*Haloxylon ammodendron*) – شجيرة تُزرع على مستوى العالم في المناطق المُعرَّضة للتصحر وتتناثر في الصحاري الرملية والكثبان غير المستقرة، جنبًا إلى جنب مع أنواع أخرى من الشجيرات الصحراوية المغصنة (التي تتحمل الجفاف) مثل نبات الأرطة أو العبل (*Calligonum comosum*) والعشب الصحراوي الأجمي (*Panicum turgidum*)، كما نضرب مثالًا آخر بشجيرة الرمث الخريزي (*Haloxylon salicornicum*) الجافة والتي تُمثِّل إحدى سمات المنطقة وتنتشر عبر صحارى شبه الجزيرة العربية. ومن بين الشجيرات القزمة كثيفة الأوراق التي تكثر في التربة الرملية وتغطي المسطحات المالحة شجيرات نبات الهرم (*Zygophyllum coccineum*) والشنان المفصلي (*Anabasis articulata*). هناك أيضًا الثليث المخروطي (*Halocnemum strobilaceum*)، وهو شجيرة صحراوية ذات قيمة ترعى عليها الماشية وتنتشر في مناطق السبخة المنتشرة في المسطحات الطينية المالحة في جنوب غرب العراق. وفي بعض المناطق، يتطفل على الشجيرات الصحراوية المذكورة أعلاه سنابل صفراء وبنفسجية عديمة الأوراق لنبات الذؤنون الأنبوبي (*Cistanche tubulosa*)، وهو ما يعطيها مظهرًا متقطعًا بعد هطول الأمطار في أواخر الشتاء.

تُعد الأجمات شبه الصحراوية موطنًا لعددٍ هائلٍ من الحوليات التي تنمو بعد هطول الأمطار شتاءً، وتشمل القتاد الصحراوي (الاستراجالوس (*Astragalus*)) مثل نبات الرُخَامَة أو بَيْض الجَمَل (*A. tribuloides*)، وأعشاب مثل الصمعاء (*Stipa capensis*) وشبيه الأقحوان، وهو أحد أشكال نبات القيامة (*Pallenis heirochuntica*).

السعدانة المفترشة أو «السعدان» (*Neurada procumbens*)، نبات حولي رعوي ينتشر عبر الكثبان الرملية للشرق الأوسط وصحاريه ويُنتج ثمرة شوكية مميزة، وقد أُستخدم النبات في أماكن أخرى كغذاءٍ للإبل، وأُستخدم كذلك في الطب العربي القديم لعلاج الإسهال والزُحار (الدوسنتاريا) (تشيلالبا وآخرون، 2020).

تتكوّن السهول الرسوبية الوسطى ومنطقة الأهوار الجنوبية في العراق من مجموعة بحيرات عذبة المياه وضحلة ومستنقعات وأهوار وسهول تغمرها المياه موسميًا بين نهري دجلة والفرات. وتكون هذه السهول الرطبة محطة مُهمة للطيور المهاجرة شتاءً في أوراسيا وواحة للتنوّع البيولوجي؛ حيث يُقدَّر عدد النباتات فيها بما يربو على 800 نوع، وهو عددٌ لا يحظى بمثيله موقعٌ آخر.

الأراضي الشجيرية المختلطة (65,000 كم²) – تضم السهوب الجافة والرطبة في شمال العراق، على الحدود مع سوريا من جهة الشرق وإيران من جهة الغرب. تتكوّن الأراضي الشجيرية في العراق من السهول العليا وحتى سفوح الجبال؛ والسهول الحصوية والنتوءات الصخرية وممتلكات صغيرة مزروعة بالمحاصيل الشتوية. وتقع منطقة الجزيرة العليا ذات الطبيعة المناخية المُحدَّدة، وقضاء كركوك، وقضاء أربيل، ومحافظة نينوى، ومنطقة سفوح التلال الفارسية ضمن هذه المنطقة. يرتفع المعدل السنوي لهطول الأمطار بالمنطقة الصحراوية ارتفاعًا كبيرًا مقارنة بمنطقة السهوب؛ حيث يتراوح بين 350–200 مم و500–350 مم. تتميز السهوب بكونها أراضٍ تنمو بها أعشاب متفرقة وقصيرة الطول وشجيرات صغيرة متناثرة. ويضم الغطاء النباتي الذي تمتاز به هذه المنطقة أعشاب المروج مثل القبأ البصيلي (*Poa bulbosa*) والسعادي رفيع الأوراق (*Carex stenophylla*) والبقاع الملونة للحوذان الآسيوي (*Ranunculus asiaticus*) وشقائق النعمان (*Anemone coronaria*) في الربيع. وتظهر التجمعات النباتية المتبقية للشيح الأبيض (*Artemisia herba-alba*) والقيصوم أو الأخِيليا (*Achillea conferta*) في منطقة السهوب، وتعد المنحدرات الصخرية موطنًا للنباتات البصيلية التي تشمل الصاصل (*Ornithogalum neurostegium*)، ودلبوث الحصاد (*Gladiolus italicus*)، ونباتات الفصيلة البلفية، والبُلْبُوس طويل الساق (*longipes Leopoldia*). تنتشر الأشجار الصغيرة مثل البطم (*Pistacia*) في الأراضي العشبية المفتوحة في السهوب الجافة، بيْد أن أراضي الرعي والأراضي الزراعية حلّت محلَّ الغطاء النباتي الطبيعي بصورة كبيرة.

الغابة الجبلية (30,000 كم²) – تشمل الجبال الشمالية المتاخمة لسوريا وتركيا وإيران؛ بما في ذلك مرتفعات جبل سنجار، وتقع ضمن المنطقة البيئية المهمة لسهوب غابات جبال زاجروس. ويُقدَّر المعدل السنوي لهطول الأمطار في المنطقة 1400–700 مم، بعضه في صورة غطاء ثلجي. تُشكِّل المنطقة غابة مفتوحة إلى مغلقة يتنوّع شكل الغطاء النباتي فيها متخذًا مظهر «الوسادة الشائكة». وتسود هذه الغابات أنواعٌ مختلفةٌ من البلوط (السنديان الرومي (*Quercus aegilops*)، والسنديان البرانتي (*Q. brantii*)، وبلوط العفص (*Q. infectoria*)، والسنديان اللبناني (*Q. libani*) تتخللها شجرة التربنتين الفارسية (*Pistacia atlantica*) وبطم كينجوك (*P. khinjuk*)، وكذلك غابات الصنوبر (الصنوبر الأفغاني (*Pinus brutia* var. *eldarica*). ويضم التجمع النباتي المتخذ شكل وسادة شائكة أرض شجيرية مفتوحة تسودها قباب كبيرة شوكية من القتاد

(أنواع الاستراجالوس (*Astragalus*)، مصحوبة بدفنة عود غار (*Daphne acuminata*)، وزهرة العسل (*Lonicera arborea*)، وكذلك شجيرات أخرى قصيرة وشائكة متنوّعة، مثل الغملول (*Acantholimon*)، وأكانثوفيلوم (*Acanthophyllum*)، وكوزنية (*Cousinia*). وقد تعرَّض الكثير من مناطق الغابة الأصلية للإزالة، ولا تظهر الغابات الخيمية المُغلقة سوى في عددٍ قليلٍ من المواقع التي يتعذّر الوصول إليها.

غابة جبال الألب (100 كم²) – تضم منطقة جبال الألب (2750–3730 م) من الجبال الشمالية بالبلاد، وتشمل قضاء العمادية وقضاء راوندوز ومحافظة السليمانية. يزيد المعدل السنوي لهطول الأمطار عن 1000 مم في هذه المناطق، ومعظمها أمطار ثلجية. تقع هذه المنطقة، شأنها شأن سابقتها، ضمن المنطقة البيئية المُهمة لسهوب غابة جبال زاجروس، لكن منحدراتها الأكثر ارتفاعًا من سابقتها تستضيف شكلًا آخر من أشكال الحياة النباتية الذي تسوده أعشاب وشجيرات مُعمرة قصيرة الطول من الفصيلة النجمية (Asteraceae)، والفصيلة الكرنبية (Brassicaceae)، والفصيلة البقولية (Fabaceae)، والفصيلة الشفوية (Lamiaceae)، والفصيلة الرصاصية (Plumbaginaceae)، والفصيلة النجيلية (Poaceae). على الرغم من صغر مساحة هذه المنطقة مقارنة بسابقاتها، فإنها تُمثِّل موطنًا لعددٍ كبيرٍ من الأنواع النباتية النادرة والمتوطنة لا يتناسب ومساحتها. وتعد الجبال الشمالية للعراق في كردستان امتدادًا لنظام جبال الألب الممتدة من البلقان شرقًا عبر جنوب تركيا وشمال العراق وإيران وأفغانستان، وصولًا إلى جبال الهيمالايا. وفي العراق، تضم الجبال الممتدة في قضاء العمادية وقضاء راوندوز ومحافظة السليمانية أكبر عددٍ من أنواع النباتات المتوطنة، ولا سيّما جبل أفرومان، الذي يمتد متخذًا شكل سلسلة جبال (2500 م) بطول الحدود الإيرانية 25–15 كم في الجهة الشمالية الشرقية وشمال حلبجة وبيره مكرون، وقمة جبلية عالية (2800 م) على بعد حوالي 30 كم شمال غرب السليمانية.

طب الأعشاب التقليدي في العراق

يرجع تاريخ العلاجات النباتية إلى نشأة الجنس البشري، وللطب التقليدي في العراق على وجه الخصوص تاريخ طويل وثري يمكن تتبعه إلى العصر السومري (1970–4000 ق.م.) والعصر البابلي والعصر الآشوري (549–1970 ق.م.) على التوالي. وتعتمد المجتمعات الحضرية والريفية في المناطق كافة من صحراء العراق وسهوبها ومرتفعاتها إلى يومنا هذا على هذا التراث المعرفي الثري (الدوري، 2014)، ويتشابه الأمر في خارج العراق أيضًا؛ فوفقًا لمنظمة الصحة العالمية، يُلبي 80% من السكان في البلدان النامية احتياجاتهم الطبية الأساسية بالاعتماد على الأدوية العشبية التقليدية. وعلى الرغم من أن الطب الحديث أصبح أكثر توفرًا في العراق، إلا أن طب الأعشاب التقليدي ما زال يُشكِّل جزءًا حيويًا في نظام الرعاية الصحية، لا سيّما في المناطق الصحراوية حيث يمارسه «العطارون» (المُعالجون بالأعشاب) في البلاد. وبالرغم من أهمية المُعالجين بالأعشاب في المجتمع، إلا أن عملهم لا يُوثَّق في أغلب الأحيان ما يُعرِّض التراث الغني لطب الأعشاب التقليدي لمخاطر الفقدان الجسيمة.

تُبيِّن الأبحاث التي تُجرى على ممارسة طب الأعشاب بواسطة المُعالجين التقليديين في محافظتي السليمانية وأربيل في كردستان أن العديد من الأنواع النباتية التي لا تقل عن ٤٣ فصيلة تُحصد في هذه المنطقة؛ حيث تُحصد الأوراق، وكذلك الزهور والبذور ولكن بدرجة أقل من الأوراق في أغلب الأحيان من فصيلة النعناع (الفصيلة الشفوية) (Lamiaceae) ، وفصيلة الجزر (الفصيلة الخيمية) (Apiaceae)، وفصيلة الأقحوان (الفصيلة النجمية) (Asteraceae) وفصيلة البازلاء والفول (الفصيلة البقولية) (Fabaceae) (أحمد، 2016؛ نقشبندي 2015). وتتمثل أكثر طرق التحضير شيوعًا في المنطقة في الاستخلاص بالإغلاء (تركيز أو استخلاص نبات)، إلا أن النباتات أُستهلكت أيضًا كخضراوات أو في صورة مسحوق. وفي كردستان، تُمثِّل العلاجات النباتية الشكل الوحيد من أشكال الدواء في القرى الجبلية النائية حيث لا تكون المعرفة بتأثيرات الأعشاب واستخداماتها حكرًا على المُعالجين بالأعشاب فحسب، كما هي الحال في الأنظمة الطبية التقليدية الأخرى، بل تُمثِّل جزءًا لا يتجزأ من التراث الثقافي للأسرة وتنتقل عبر الأجيال. وتضم جميع الأسواق والمتاجر قسمًا تُباع فيه النباتات البرية الطبية، مثل بصيلات الخزامى البرية الطازجة كمُسكِّنٍ تقليديٍّ للألم (أمين وآخرون، 2016). وتبيع متاجر العلاج بالأعشاب في المنطقة علاجات نباتية من مصادر برية ومزروعة من جميع أنحاء الشرق الأوسط وآسيا الصغرى، وعلى الرغم من أن معظم النباتات المتداولة قد أُستخدمت منذ القدم، إلا أن المُعالجين بالأعشاب يشيرون في عام 2011 إلى الازدياد المُطرد في التجارة بسبب الاستقرار الاقتصادي وقد أُستورد ما يصل إلى 64% من النباتات الطبية من خارج العراق في ذلك الوقت (ماتي ودي بوير، 2011). جديرٌ بالذكر أن النباتات في العراق موثقة أيضًا لاستخدامات تقليدية أخرى بخلاف استخدامات طب الأعشاب؛ بما في ذلك: الأغذية، والأدوات، والعلكات، والأعلاف، والدباغة، والأصباغ (أحمد وعسكري، 2015).

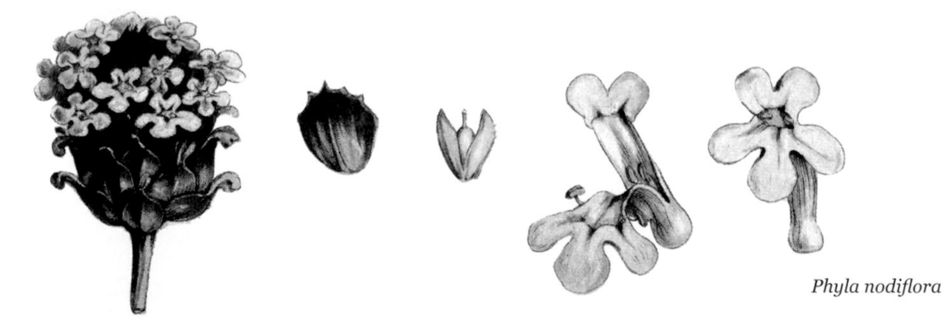

Phyla nodiflora

ABOUT THIS BOOK

THE SELECTED PLANTS

The plants depicted in this herbal flora were inspired by the work of Abdul Jaleel Ibrahim Al-Quragheely in Iraq, and augmented by other reports published in the literature. Most of the species described are native and reported to be used widely. Some, such as ginger (*Zingiber officinale*) were introduced to the region but feature prominently in Islamic medicine, so are also included. We must emphasize that few of the plant species reported to have medicinal benefits (or indeed preparations containing them) have been evaluated scientifically for their efficacy and safety. For this reason, herbal medicine should always be applied judiciously, with due care and experience.

Many plants are poisonous, even in small doses. Ensuring that herbal medicine is practised safely is a challenge that exists beyond Iraq. It requires rigorous plant identification carried out by knowledgeable herbalists and regulated procedures relating to the handling, storage and application of the medicinal plants (Abu-Irmailleh & Afifi, 2003). This herbal flora is intended to serve as a visual documentation and celebration of the traditional use of plants in herbal medicine in Iraq, inspired by the work carried out in a particular community. It is not a clinical guide to the medicinal properties of plants, or a treatise on how to administer them. Finally, with climate change and shifts in land use, many plants in Iraq face decline or even extinction in the future. No plant should be harvested in a way that is unsustainable, and rare or endangered species should be left alone completely.

THE STORY OF THE DOCTOR OF AL-ASHAB

Abdul Jaleel Ibrahim Al-Quragheely was born in 1934 in Baghdad. After graduating from the College of Agriculture Baghdad in 1959 he was appointed to Iraq's Ministry of Higher Education as a first agricultural engineer, where he remained until 1980. Together with his wife, Nazik Saeed Al-Jarrah, a biology teacher, he had a family of seven children, who shared a passion for natural history with their parents. Following his appointment at the Ministry of Higher Education, Abdul Jaleel Ibrahim ran a 40 ha (100 acre) farm near Radwaniyah in central Iraq where he cultivated plants, including medicinal herbs. In 1988, Abdul Jaleel was

diagnosed with diabetes, which he treated himself using medicinal herbs grown on his farm. So began his idea of developing a community practice in traditional herbal medicine. He dedicated a part of his house to the preparation of medicinal herbs grown on the farm. The practice grew in popularity and importance, eventually serving the whole of the Karkh region in Baghdad. He called it *Al-Ashab*.

Known as 'the Doctor of Al-Ashab', Abdul Jaleel's work in delivering herbal medicine to the community was widely appreciated in Iraq. He shared his findings widely, for example speaking about his findings on the treatment of diabetes using herbal medicine at the Al-Yarmouk Medical Conference. Tragically, during the invasion of Baghdad in 2003, Al-Ashab was targeted by an explosion. Much of Abdul Jaleel's work including meticulous records, documents, and plants were destroyed along with the practice. But, committed to practising herbal medicine, Abdul Jaleel continued working with communities to dispense plant-based remedies in Mansour, Baghdad for a further six years at a time when communities were reliant on plants for healthcare.

By documenting the knowledge he gained while working with local communities, Abdul Jaleel compiled a comprehensive herbal flora of Iraq in three volumes. This contained a wealth of information on hundreds of plants and their traditional use in remedies. Tragically, however, in 2009, Abdul Jaleel was killed during a terrorist attack on Al-Yousifia. His flora, and its wealth of knowledge and cultural heritage, survive. They were bequeathed to his daughter Rana in 2011 in Syria – the only safe meeting point at that time – by her mother for safekeeping in Oxford. This herbal flora is the product of Abdul Jaleel's work and devotion to plant-based medicine in Iraq.

Eminium spiculatum

نبذة عن هذا الكتاب

النباتات المُختارة

النباتات المُصوَّرة في كتاب النباتات الذي بين أيدينا مستوحاة من عمل عبد الجليل إبراهيم القره غولي في العراق وقد عززناها بتقارير أخرى منشورة في أعمال مطبوعة. معظم الأنواع المشار إليها في هذا العمل أنواع محلية ومن الشائع استخدامها. وقد استُقدِمَت بعض هذه الأنواع، مثل الزنجبيل (*Zingiber officinale*) إلى المنطقة، ولكن لها مكانة بارزة في الطب الإسلامي، ومن ثمَّ أوردناها أيضًا في هذا الكتاب. ينبغي لنا أن نوضح أن القليل من أنواع النباتات التي وردت تقارير عن فوائدها الطبية (أو المستحضرات التي تحتوي عليها بالفعل) ثبتت فعاليتها وأمانها علميًا؛ ولهذا السبب يجب دائمًا اتّباع الحكمة وتوخي العناية والاستفادة من الخبرة عند استخدام طب الأعشاب.

هناك الكثير من النباتات السامة، حتى وإن تم تعاطيها بجرعات صغيرة. ويُمثِّل التأكد من عنصر السلامة في ممارسة طب الأعشاب تحديًا يتجاوز حدود العراق؛ حيث يستلزم من المُعالجين بالأعشاب واسعي الاطلاع التعرّف على النباتات على نحو دقيق، فضلًا عن اتّباع إجراءات منظمة تتصل بمناولة النباتات الطبية وتخزينها واستخدامها (أبو الرميلة وعفيفي، ٢٠٠٣). ومن المستهدف أن يكون كتاب النباتات العشبية الماثل وثيقة مرئية وتتويجًا لاستخدام النباتات في طب الأعشاب في العراق المستوحى من العمل المنفذ في مجتمعٍ مُعينٍ، ولا يهدف هذا الكتاب أن يكون دليلًا طبيًا للخصائص الطبية للنباتات أو كيفية تعاطيها. وختامًا، فإنه في ظل التغير المناخي والاختلافات في كيفية الاستفادة من الأراضي، تواجه الكثير من النباتات في العراق خطر الاندثار، أو حتى الانقراض مستقبلًا. ومن ثمَّ لا بدّ من اجتناب استخدام الطرق غير المستدامة في حصاد أي نباتات، علاوةً على أنه يلزم الامتناع تمامًا عن حصاد الأنواع النادرة أو المُهددة بالانقراض.

قصة الدكتور العشَّاب

ولد عبد الجليل إبراهيم القره غولي في بغداد عام ١٩٣٤ وعُيِّن مهندسًا زراعيًا أول بوزارة التعليم العالي في العراق بعد تخرجه في كلية الزراعة ببغداد عام ١٩٥٩ حيث استمر عمله بهذا المنصب حتى١٩٨٠. وقد تزوج بنازك سعيد الجرَّاح، مُدرسة علم الأحياء، ورُزق منها بسبعة من الأبناء، وشاركته زوجته شغفهما بالتاريخ الطبيعي. وفي أعقاب تعيينه بوزارة التعليم العالي، أدار عبد الجليل إبراهيم القره غولي مزرعة ممتدة على مساحة ١٠٠ فدان بالقرب من الرضوانية بوسط العراق حيث زرع النباتات ومن ضمنها الأعشاب الطبية. وفي عام ١٩٨٨ شُخصت حالة عبد الجليل كمريضٍ بداء السكري، فعالج نفسه بنفسه باستخدام أعشاب طبية زرعها في مزرعته الخاصة، ومن هنا انبثقت فكرته بشأن تطوير ممارسة طب الأعشاب التقليدي على نطاق مجتمعي. وقد خصص عبد الجليل جزءًا من منزله لتحضير الأعشاب الطبية المزروعة بمزرعته، وذاعت شهرة عيادته وأهميتها والتي صارت في النهاية تخدم منطقة الكرخ في بغداد بأكملها وأطلق عليها عبد الجليل اسم «العشَّاب».

حظى عمل عبد الجليل الذي عُرف باسم «الدكتور العشّاب» الذي استهدف توصيل طب الأعشاب للمجتمع بتقديرٍ بالغٍ في العراق؛ حيث شارك مكتشفاته على نطاقٍ واسعٍ، مثلما فعل عندما تحدّث عن اكتشافاته في مجال علاج داء السكري باستخدام طب الأعشاب في مؤتمر اليرموك الطبي. وقد استُهدفت عيادة العشّاب في تفجيرٍ مأساويٍّ إبان احتلال بغداد في عام ٢٠٠٣، ما أسفر عن تلف الكثير من أعماله؛ من بينها تسجيلات ووثائق شديدة الدقة ونباتات، وأدى إلى تدمير العيادة نفسها. واصل عبد الجليل عمله المجتمعي في وصف العلاجات النباتية في منصور ببغداد مدفوعًا بالتزامه تجاه ممارسة طب الأعشاب حيث عمل لمدة تزيد عن ستة أعوام في وقتٍ اعتمدت فيه المجتمعات على النباتات في الرعاية الصحية.

في الوقت الذي كان عبد الجليل إبراهيم القره غولي يُحصِّل المعارف التي اكتسبها من العمل في المجتمعات المحلية، جمع مجموعة شاملة من النباتات العشبية التي تنبت في العراق في ثلاثة مجلدات ضمت ثروة من المعلومات عن مئات النباتات واستخداماتها التقليدية في العلاج. وقد لاقى عبد الجليل حتفه في حادثة مأساوية عام ٢٠٠٩ إثر هجوم إرهابي استهدف اليوسفية، بينما ظلت نباتاته وثروته من التراث المعرفي والثقافي على قيد الحياة؛ حيث تسلّمتها ابنته رنا عام ٢٠١١ في سوريا – نقطة الالتقاء الآمنة الوحيدة في هذا الوقت – من أمها لحفظها في أكسفورد؛ فهذه النباتات العشبية هي حصيلة عمله وتفانيه في مجال الطب النباتي في العراق.

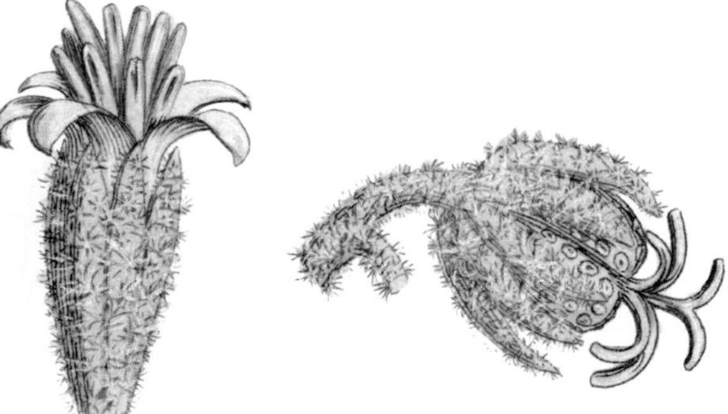

Chrozophora tinctoria

THE PLANTS

For each plant we have provided the scientific and common names, a short description, and, brief notes on distribution, ecology and the typical months of flowering. All vernacular names are in Arabic, used in Iraq and the wider Arab world.

Glaucium corniculatum

النباتات

أوردنا في هذا الكتاب الاسم العلمي لكل نباتٍ من النباتات المُشار إليها، وكذا اسمه الشائع، ووصفه الموجز، وتوزيعه، وملاحظات موجزة بشأن علاقته بالبيئة المحيطة به، ومواسم إزهاره. جميع الأسماء الدارجة المُستخدمة في الكتاب هي الأسماء العربية المتداولة في العراق والعالم العربي على نحو أشمل.

Agrimonia eupatoria L.

Family Rosaceae

الفصيلة الوردية

common agrimony

ghafat sha', غافث شائع

An erect perennial to 1.2 m (4 ft) with a woody rootstock. Stems hairy. Leaves compound, leaflets narrowly obovate with serrated margins. Flowers yellow, campanulate, shortly pedicellate, in spike-like racemes. Stamens many. Fruit strongly deflexed, 1–2-seeded.

Common agrimony is native to Macaronesia, Europe to Afghanistan and northwest Africa. In Iraq it is common in the forest zone, on hillsides, often in cool, shaded habitats and cultivated fields; alt. 700–1,800 m (2,300–5,900 ft); fl. June–Sept.

Common agrimony has been known as a medicinal herb since ancient times. It has been used to treat weak memory, eye disorders and liver diseases. Its therapeutic properties extend to its use as an anti-diarrhoeal, as an astringent in the treatment of external wounds and as a mild diuretic.

In Iraq, the aerial parts have been used to stop bleeding and to heal wounds. It has been used as a digestive tonic, to treat sore throat, diarrhoea, cystitis and urinary incontinence, and for the treatment of kidney stones. Al-Douri (2000) notes that a poultice of aerial parts has been used for skin diseases and against haemorrhoids. A decoction of the aerial parts has been used as a tonic, diuretic, anthelmintic and for jaundice.

نبات منتصب مُعمر يبلغ طوله 2.1 م ذو جذر خشبي. ساق النبات مكسوة بشعيرات، وأوراقه مركبة، ووريقاته تتخذ شكلًا بيضاويًا مقلوبًا يضيق من موضع عنق الوريقة وحوافها مسننة. النبات له أزهار صفراء جرسية معنقة متراصة في شكل مجموعات عنقودية تشبه السنابل. النبات له أسدية كثيرة. الثمرة تنمو بصورة منحنية وتحتوي على بذرة واحدة أو بذرتين.

الموطن الأصلي لنبات غافث شائع هو ماكارونيسيا، والمنطقة من أوروبا إلى أفغانستان، وشمال غرب أفريقيا. ويشيع نمو نبات غافث في العراق في منطقة الغابات وعلى جوانب التلال في الموائل الباردة الظليلة والحقول المزروعة في غالب الأمر. ينمو النبات على ارتفاع 700–1800 م؛ موسم الإزهار: يونيو-سبتمبر.

عُرف الغافث الشائع بأنه عشب طبي منذ القدم، وقد استُخدم في علاج ضعف الذاكرة واضطرابات العين وأمراض الكبد. تمتد خصائصه العلاجية إلى استخدامه كمضادٍ للإسهال وكمادة قابضة في علاج الجروح الخارجية وكمُدرٍّ خفيفٍ للبول.

في العراق، استُخدمت الأجزاء الهوائية من النبات لوقف النزيف وتضميد الجروح، كما استُخدم الغافث الشائع كمنشطٍ للجهاز الهضمي، وفي علاج التهاب الحلق والإسهال والتهاب المثانة وسلس البول وحصوات الكلى. يشير الدوري (2000) إلى استخدام الأجزاء الهوائية من النبات ككماداتٍ في الأمراض الجلدية، ووضعها على البواسير. يعمل المُستخلص من الأجزاء الهوائية للنبات بالإغلاء كمُنشطٍ ومُدرٍّ للبول ومضادٍ للديدان وعلاجٍ لليرقان.

Alhagi maurorum Medik.

Family Fabaceae
(Leguminosae: Papilionoideae)

الفصيلة البقولية
(القرنيات: الفراشيات)

camel thorn; manna plant

nabat al man نبات المن؛ ; *aqul* عاقول; *a'gul* عاكول; *al hag* الحج; *shauk al jamal* شوك الجمل; *hushtirālūk* حشترالوك (Kurd.)

A small perennial shrub to 1 m (3.3 ft), with erect to ascending branches; stems and lateral branches with stout axillary spines. Leaves simple, about 20 mm (0.8 in) long, obovate. Flowers pea-like, red or purplish-red, about 12 mm (0.5 in) long, borne in axillary racemes. Pods brown-black when mature, 10–30 mm (0.4–1.2 in) long, linear, constricted between the seeds.

Camel thorn is native to southeast Greece to Siberia and north-central India. In Iraq it is found in lower mountain valleys, degraded grassland, margins of springs, on saline soils in desert, in wasteland by fields, ditches and canal banks, often on slightly saline soils; alt. 0–1,500 m (4,900 ft); fl. June–Aug.; fr. Aug.–Dec.

In Iraq this plant has been used as an expectorant, a mild laxative and a diuretic. It has also been used for treating rheumatism, and the flowers to stop bleeding. In the Arabian Peninsula, an infusion of the root is boiled in water with lemon; the whole plant has been used for treating cataracts, jaundice, migraine, painful joints and as an aphrodisiac.

Camel thorn is also known as the manna plant, with reference to the *man* in the Bible and of Semitic languages in general. The plant emits an exudate apparently in the absence of insects, in the form of brownish tear-shaped drops; it is considered one of the most economically important products in Iran (called *taranjabin* Pers.). It is used in traditional medicine as a laxative and expectorant, and in the preparation of special sweetmeats. Ibn al-Baitar (c. 1240 CE) mentions that this plant was very common in Iraq and that its syrup was used for leucoma, and that a salve made from it known as *burūd* was used to treat impairment of vision.

شجيرة صغيرة مُعمرة يبلغ طولها 1 م، ذات فروع منتصبة ومتسلقة. سوق النبات وفروعه الجانبية ذات أشواك إبطية سميكة. الأوراق بسيطة ويبلغ طولها نحو20 مم وبيضاوية مقلوبة. الأزهار تشبه البازلاء، وتتراوح بين اللون الأحمر والأرجواني ويبلغ طولها نحو12 مم، وتكون محمولة على عناقيد إبطية. عندما ينضج النبات تنمو بين البذور أجربة خطية متقلصة بنية-سوداء اللون يصل طولها إلى 30–10 مم.

شوك الجمل موطنه جنوب شرق اليونان إلى سيبيريا وشمال وسط الهند. يوجد في العراق في الوديان الجبلية المنخفضة، والمراعي المدرجة، وعلى أطراف الينابيع، وفي التربة المالحة في الصحراء، وفي الأراضي القاحلة بالقرب من الحقول، والخنادق، وضفاف القنوات. غالبًا ما ينمو النبات في التربة قليلة الملوحة، على ارتفاع 1500–0 م؛ موسم الإزهار: يونيو – أغسطس؛ موسم الإثمار: أغسطس - ديسمبر.

في العراق، استُخدم النبات كمقشعٍ ومُلينٍ خفيفٍ وكمُدرٍّ للبول، كما استُخدم لعلاج الروماتيزم. واستُخدمت الأزهار لوقف النزيف؛ حيث يتم في شبه الجزيرة العربية غلي نقيع الجذر في الماء مع الليمون، فيما استخدم النبات بأكمله كعلاج لإعتام عدسة العين واليرقان والصداع النصفي (الشقيقة) وآلام المفاصل وكمُنشطٍ جنسيٍّ.

يُعرف شوك الجمل أيضًا باسم نبات المن، إشارة إلى المن في الكتاب المقدس وفي اللغات السامية بوجهٍ عامٍ. يُفرز النبات سائلًا ظاهرًا على شكل قطرات دمعية بنية اللون في غياب الحشرات. ويعد شوك الجمل واحدًا من المنتجات الإيرانية ذات الأهمية الاقتصادية (ويُطلق عليه بالفارسية اسم ترنجبين. استُخدم النبات في الطب القديم كمُلينٍ وطاردٍ للبلغم وفي إعداد بعض أنواع الحلوى. ذكر ابن البيطار (نحو عام 1250 م) أنه قد شاع هذا النبات في العراق وأن شرابه استُخدم لعلاج الغُفاءة (عَتَامَةٌ على القرنيَّة)، كما صُنع منه مرهمٌ عُرف باسم برود برود كان يُستخدم في علاج ضعف البصر.

Althaea officinalis L.

Family Malvaceae

marshmallow

الفصيلة الخبازية

خطمي; نبات الخطمي; *khatmī* خطمي;
khatmiya خطمية; *ghasūl* غاسول

An erect perennial to 1 m (3.3 ft); root thick and fleshy. The whole plant has white stellate hairs. Lower leaves round to somewhat trilobed; upper leaves 3–5-lobed, 30–50 mm (1.2–2 in) long; all leaves with irregularly toothed margins. Flowers 20–30 mm (0.8–1.2 in) across, pale pink, single or paired in the axils of leaves. Fruit dry, many-segmented (carpellate), partially enveloped by the sepals, each segment with a kidney-shaped to rounded seed.

Marshmallow is native from Europe to central Siberia and Pakistan, and northwest Africa. In Iraq it is recorded as very rare, found only in the lower forest zone at about 700 m (2,300 ft); fl. June and July.

Marshmallow has been esteemed for its medicinal properties since ancient times, the leaves being used as a poultice and mixed with oil applied to soothe burns, insect bites and stings. An infusion of the root was given for coughs and for digestive problems. The flowers yield a red dye, and glue can be obtained from its roots.

In Iraq it has been used to treat digestive problems, including irritable bowel syndrome; the leaves have been recommended for treating cystitis and urinary incontinence. The plant has also been used for the relief of dry coughs, asthma and bronchitis; an infusion of the flowers or dried flower powder to treat skin inflammation; and ointment made from the roots for skin pimples and abscesses, and also as a mouthwash to relieve inflammation of the gums. Peeled roots have been recommended as lollipops for children to help aid the growth of teeth.

نبات مُعمر منتصب يبلغ طوله 1 م. الجذور سميكة وغليظة. النبات بالكامل مُغطى بشعيرات نجمية بيضاء اللون. الأوراق السفلية مستديرة إلى ثلاثية الفصوص إلى حد ما، أما الأوراق العلوية فهي مكونة من 3-5 فصوص ويبلغ طولها 50–30 مم. جميع أوراق النبات ذات حواف مسننة غير منتظمة الشكل. الأزهار وردية باهتة، ومفردة أو مقترنة عند محاور الأوراق ويبلغ عرضها 30–20 مم. الثمار جافة، ذات أخبية كثيرة مجزأة (مكربلة)، يلفها الكأس جزئيًا. كل كربلة لها بذرة تشبه حبة الفول في شكلها أو تتخذ شكلًا مستديرًا.

نبات الخطمي موطنه الأصلي من أوروبا إلى وسط سيبيريا وباكستان وشمال غرب أفريقيا. سُجّل النبات في العراق ضمن النباتات شديدة الندرة؛ حيث لا يوجد سوى في منطقة الغابات المنخفضة على ارتفاع حوالي ٧٠٠ متر؛ موسم الإزهار: يونيو ويوليو.

يُستخدم نبات الخطمي منذ القدم لما له من خصائص طبية. وقد استُخدمت الأوراق ككمادة ومزجت بالزيت لتهدئة الحروق ولدغات الحشرات ولسعاتها. يُستخدم منقوع الجذور لعلاج السعال واعتلالات الجهاز الهضمي. تُنتج الأزهار صبغة حمراء ويمكن الحصول على غراء من جذور النبات.

استُخدم نبات الخطمي في العراق لعلاج اعتلالات الجهاز الهضمي بما في ذلك متلازمة القولون العصبي. يُنصح باستخدام أوراق النبات لعلاج التهاب المثانة وسلس البول. استُخدم النبات لتهدئة السعال الجاف وحالات الربو والتهاب الشعب الهوائية، كما استُخدم منقوع الأزهار أو مسحوق الأزهار المجففة لعلاج التهاب الجلد. وقد استُخدمت جذور النبات في تحضير مرهمٍ لعلاج البثور والخراجات الجلدية، كما استُخدمت كغسولٍ للفم لتخفيف التهاب اللثة. يُوصى باستخدام الجذور المقشرة كمصاصات للأطفال للمساعدة في نموّ الأسنان.

Ammannia baccifera L.

Family Lythraceae

الفصيلة الخثرية

blistering ammannia

rigl hamāmah رجل حمامة

An annual or perennial herb, 20–70 cm (8–28 in) tall. Stems branched, 4-angled to narrowly winged above. Leaves opposite, narrow lanceolate, with a clasping base and entire margins. Flowers small, without petals, borne in dense clusters in the axils of leaves. Fruit a capsule, small to about 2 mm (0.008 in).

Ammannia baccifera is native to tropical and North Africa, the Arabian Peninsula east through Iran and India to Malaysia; also found in Australia and the Caribbean Islands. In Iraq it is common in the central and eastern regions, especially in wet and marshy places, rice fields and on river banks; alt. 10–20 m (33–66 ft); fl. Aug.–Dec.

The leaves have been used in traditional medicine to soothe irritation and inflammation of the skin and applied to open septic abscesses.

عشبة حولية أو مُعمرة، يبلغ طولها 20–70 سم. الساق متفرعة ذات 4 زوايا ومُجنحة ضيقة من أعلى. الأوراق متقابلة رمحية ضيقة ذات قاعدة قابضة وحواف كاملة. الأزهار صغيرة خالية من البتلات، ومُحملة في مجموعات كثيفة على محاور الأوراق. الثمرة جرابية صغيرة تصل إلى 2 مم.

الموطن الأصلي لنبات رجل حمامة في أفريقيا الاستوائية وشمال أفريقيا وشرق شبه الجزيرة العربية عبر إيران والهند إلى ماليزيا، كما يوجد النبات في أستراليا وجزر الكاريبي. يشيع وجود النبات في المنطقة الوسطى والمنطقة الشرقية، لا سيّما في الأماكن الرطبة والأهوار ومزارع الأرز وعلى ضفاف الأنهار. ينمو النبات على ارتفاع 10–20 م؛ موسم الإزهار: أغسطس – ديسمبر.

استُخدمت الأوراق في الطب التقليدي لتهدئة تهيّج البشرة والتهابها وتوضع على الخراجات الإنتانية المفتوحة.

Anastatica hierochuntica L.

Family Brassicaceae

الفصيلة الكرنبية

rose of Jericho

kaff Mariyam كف مريم

An annual herb with prostrate to ascending stems branching from the base, to 15 cm (6 in), the whole plant covered with stellate hairs. Dry branches stiffening and incurving in fruit, resembling a clenched fist. Leaves obovate to oblanceolate. Flowers minute, 2 mm (0.008 in) across, white, borne in the axils of leaves. Fruit (a silicula), ovoid-globose about 5 mm (0.2 in) across.

Rose of Jericho is native to the Sahara eastwards to Iraq and the Arabian Peninsula, and from southern Iran to southwest Pakistan. In Iraq it is found in the desert regions, on sandy soils, in wadis or in open deserts, especially in depressions where rainwater collects; alt. up to 250 m (850 ft); fl. Feb.–Apr.

There are many folk legends associated with this plant; it is believed that the Virgin Mary clenched it in her hand when giving birth to Jesus. The name *kaff Mariyam* is cited by Ibn al-Baitar (c. 1240 CE) together with several other names used in the Middle East. Reported by Mati & de Boer (2011) to bring luck to pregnant women – because the plant unfurls when soaked in water – it is said that the woman will give birth successfully. In the Arabian Peninsula, the dried plant is reported to have been soaked in water and the solution drunk by women during childbirth.

عشبة حولية ذات ساق زاحفة إلى صاعدة تتفرع من القاعدة بطول 51 سم. النبات بأكمله مغطى بشعيرات نجمية الشكل. الفروع جافة متصلبة ومنحنية إلى الداخل حاملة الثمار في شكل يشبه قبضة يد محكمة الإغلاق. الأوراق بيضاوية مقلوبة إلى مطوية. الأزهار دقيقة عرضها 2 مم وبيضاء اللون ومُحملة على محاور الأوراق. الثمرة (خردلة) بيضاوية الشكل يصل عرضها إلى 5 مم.

ورد أريحا موطنها في الصحراء الممتدة شرق العراق وشبه الجزيرة العربية، ومن جنوب إيران إلى جنوب غرب باكستان. توجد النبتة في المناطق الصحراوية بالعراق وتنمو في البيئة الرملية والوديان والصحارى المفتوحة، لا سيّما في المنخفضات التي تتجمع فيها مياه الأمطار. تنمو النبتة على ارتفاع 052 م؛ موسم الإزهار: فبراير– إبريل.

هناك أساطير شعبية كثيرة مرتبطة بهذا النبات؛ حيث يُعتقد أن السيدة مريم العذراء أحكمت قبضتها عليه عند ولادة السيد المسيح. استوحى ابن البيطار (نحو عام 1240 م) اسم كف مريم، بينما استُخدمت أسماء عدة في الشرق الأوسط للإشارة للنبات. أشار كلٌ من ماتي ودي بوير إلى النبات بوصفه جالبًا للحظ للمرأة الحامل لأنه يتفتح عند نقعه في الماء، ومن ثمَّ قيل أنه يُتبرك به كي تلد المرأة دونما مشكلة. يُنقع النبات في الماء، في شبه الجزيرة العربية، وتشرب النساء الماء أثناء الولادة.

Anchusa azurea Mill.

Family Boraginaceae

الفصيلة الحمحمية

bugloss

lisan al thor لسان الثور;
ward lisan al thor ورد لسان الثور;
zwan زوان; *qola zwan* قولا زوان

A coarsely hairy perennial herb, 20–95 cm (8–37 in) tall. Stems leafy. Leaves linear-elliptic to lanceolate or oblanceolate with entire to weakly crenate margins. Flowers blue, borne in branched inflorescences; corolla with a small tube with stamens inserted near the top of the tube. Fruit (nutlets), rugose, with a thick basal ring.

Bugloss is native from east-central Europe to the Mediterranean and western Himalaya. In Iraq, it is found mainly in the lower forest zone in the northeast, especially in irrigated cereal fields, along roadsides, waste land near gardens, stony ground, and by springs; alt. 100–1,700 m (390–5,600 ft); fl. Mar.–Sept.

Bugloss flowers infused in tea have been used as a tonic for invalids and children, to reduce fevers, as a sedative and as a diuretic.

عشبة مُعمرة ذات شعر خشن يصل طولها إلى 95–20 سم. الساق مورقة. الأوراق طولية - بيضاوية الشكل إلى رمحية أو مطوية. حواف الأوراق بالكامل على شكل أسنان مستديرة ضعيفة. الأزهار زرقاء اللون، محمولة في نورات متفرعة. التويجات عبارة عن أنبوب صغير به أسدية بالقرب من أعلى الأنبوب. الثمرة (تشبه الجوز الصغير) مُجعدة ذات حلقة قاعدية سميكة.

الموطن الأصلي لعشبة لسان الثور من شرق أوروبا الوسطى إلى البحر الأبيض المتوسط وغرب جبال الهيمالايا. تنمو عشبة لسان الثور في العراق بشكل رئيسي في منطقة الغابات المنخفضة في الشمال الشرقي للبلاد، لا سيّما في حقول الحبوب المروية على جانبي الطرق، وفي الأراضي القاحلة بالقرب من الحدائق والأراضي الصخرية وبجانب الينابيع. ينمو العشب على ارتفاع 1700–100 م؛ موسم الإزهار: مارس - سبتمبر.

استُخدمت أزهار لسان الثور المنقوعة في الشاي كمُقوٍ للمعاقين أو المُصابين والأطفال ولتهدئة الحمى وكمُهدئٍ ومُدرٍ للبول.

Anethum graveolens L.

Family Apiaceae

dill

الفصيلة الخيمية

الشبت; *sazāb al barr* سزاب البر ;
habbah helwah حبة حلوة

An erect, glabrous, yellowish- or bluish-green annual herb to 90 cm (35 in) tall. Stem and branches finely ridged. Leaves sheathing at base, divided into linear segments. Flowers minute, greenish-white, in many-rayed umbels. Fruit elliptic, tapering into a small crenulate-margined disc.

Dill is widely cultivated and often naturalised in the Mediterranean Basin and southeast Europe: Cyprus, Syria, Lebanon, Israel and Palestine, Jordan, Egypt, Arabia, Turkey, Iran and adjacent regions east to China, North Africa (Libya) and sporadically elsewhere. In Iraq it is common on the alluvial plains in the desert region, less so in the lower forest zone. It is widely cultivated and often sub-spontaneous in appearance as a weed in gardens, orchards, fields and waste land near villages; alt. up to 700 m (2,300 ft) or more; fl. & fr. Mar.–May on the plains, June–July in the mountains.

Dill was called ŠIBITTU (Akkadian) in ancient Mesopotamia and was recognised for its medicinal properties. In Iraq, it is widely cultivated for its foliage, which is cooked as a vegetable and used to flavour rice. The dried leaves were sold in local markets as sbint; while the ripe fruits were sold as habbat halwa (sweet seeds) and used as a condiment with cooked meats. Another name for this plant in the north is *sibat* (Kurd., which is used as a condiment for flavouring).

A medicinal oil (*Oleum anethi*) can be obtained from the fruit by distillation, its properties being carminative and stomachic. It is used to relieve flatulence and help in digestion. The oil is recognised locally, as well as elsewhere, as the basis for an excellent remedy for children's complaints such as minor digestive ailments, flatulence etc., and forms the basis of what is known as 'dill water' or 'gripe water'. Chewing the seeds improves mouth odour. The plant is good for coughs, colds and influenza. The seeds have been used to increase milk production in nursing mothers.

عشبة حولية منتصبة ملساء صفراء أو خضراء مائلة إلى الزرقة يصل طولها إلى 90 سم. الساق والفروع مسننة. الأوراق مغلفة من عند القاعدة ومُقسمة إلى قطاعات طولية. الأزهار كثيرة الفروع خيمية ودقيقة الحجم وبيضاء اللون مخضرة. الثمرة بيضاوية الشكل مستدقة الطرف تنتهي بقرص صغير ذي حواف متعرجة.

يُزرع الشبت على نطاق واسع ويُستوطن في غالب الأمر في حوض البحر المتوسط وجنوب شرق أوروبا: قبرص وسوريا ولبنان وفلسطين والأردن ومصر والمملكة العربية السعودية وتركيا وإيران والمناطق المتاخمة لشرق الصين وشمال أفريقيا (ليبيا)، ويُزرع على نحو غير منتظم في أماكن أخرى. تشيع زراعة الشبت في العراق في السهول الرسوبية بمنطقة الصحراء وتقل زراعته في منطقة الغابة المنخفضة. تكثر زراعة الشبت وينمو كحشيشة على نحو شبه تلقائي في الحدائق والبساتين والحقول والأراضي القاحلة بالقرب من القرى. ينمو الشبت على ارتفاع 700 م أو أكثر؛ موسم الإزهار والإثمار: مارس – مايو في السهول، ويونيو – يوليو على الجبال.

في اللغة الأكادية المُتحدَّث بها ŠIBITTU عُرف الشبت باسم قديمًا في بلاد ما بين النهرين واستُخدم لما له من خصائص طبية. يُزرع الشبت على نطاق واسع بالعراق حيث تُطهى أوراقه كخضراوات وتُستخدم لتنكيه الأرز. تُباع الأوراق المجففة في الأسواق المحلية باسم سبنت (*sbint*)، بينما تباع الثمار الطازجة باسم الحبة الحلوة (*habbat halwah*) وتستخدم كبهارٍ عند طهو اللحم. يُشار إلى النبات باسم سبت (*sibat*) أيضًا (ويستخدم في كردستان كبهارٍ لإضافة نكهة).

يمكن الحصول على زيتٍ طبيٍّ (زيت الشبت) من الثمار بالتقطير، وله خصائص طاردة للغازات ومُعالِجة للمغص المعدي. يُستخدم زيت الشبت لعلاج الانتفاخ والمساعدة على الهضم. عُرف زيت الشبت على النطاق المحلي وغيره كعلاجٍ ممتازٍ لشكاوى الأطفال مثل متاعب الجهاز الهضمي الخفيفة وانتفاخ البطن وما إلى ذلك واعتُبر أساسًا لتحضير ما يُعرف باسم «ماء الشبت» أو «ماء غريب». تُمضغ بذور الشبت لتحسين رائحة الفم. يُعد النبات علاجًا جيدًا للسعال ونزلات البرد والإنفلونزا، كما تُستخدم البذور لإدرار اللبن لدى الأمهات المرضعات.

Arctium lappa L.

Family Asteraceae

الفصيلة النجمية

burdock

waisar ويسا ; *arqityon* ارقتبون ;
'urqah عرقة (Kurd.)

An erect biennial, 75 cm–1 m (30–39 in) tall with widely branching stems. Stems furrowed, often somewhat purplish. Leaves grey-green, hairy beneath with prominent veins, cordate-ovate, margins with shallow broad teeth. Flowers purple in lax to rather dense, corymbose, spiny inflorescences. Fruit (achenes) compressed-obovate, brownish, striate, pappus falling.

Burdock is found almost throughout Europe, Cyprus, Lebanon, Turkey, Caucasia, Central Asia, Iran and Afghanistan east to China. It was introduced in North America and elsewhere. In Iraq it is widespread in the mountain regions in shady places, in orchards under apple or *Populus* trees or on mountains under oak or walnut, usually near streams and in other humid areas; alt. 700–1,550 m (2,300–5,100 ft); fl. & fr. June–Sept.

Burdock has a long history of use in herbal medicine. Fresh or dried roots, leaves and seeds are all used medicinally. The root has been used as a diuretic, as a poison antidote and a blood purifier. It has also been used to treat joint inflammation and chronic dermatitis and for the treatment of acne, eczema and psoriasis. Seeds have been used in fevers and infections such as mumps and measles.

نبات منتصب حوْول (يفصل بين موسم الإثمار وموسم الإثمار التالي عامان). يصل طوله إلى ٧٥ سم – ١ م. الساق كثيرة التفرع ومُخددة وأرجوانية اللون إلى حدٍ ما في غالب الأمر. الأوراق رمادية مائلة إلى الخضرة ومُشعرة من الأسفل وذات عروق بارزة، قلبية الشكل بيضاوية، وحوافها ذات أسنان عريضة قصيرة الطول. الأزهار أرجوانية اللون متراخية إلى نورات كثيفة نسبيًا عذقية شوكية. الثمرة (الأوجين) مضغوطة بيضاوية مقلوبة ومائلة للون البني ومُحززة وذات زغب متساقط.

يوجد الأرقِطيون في كل أنحاء أوروبا وقبرص ولبنان وتركيا والقوقاز ووسط آسيا وإيران وأفغانستان في الجهة الشرقية من الصين تقريبًا، كما استُقدم الأرقِطيون إلى أمريكا الشمالية وغيرها من البلدان. انتشر النبات في العراق في البقاع الظليلة من المناطق الجبلية وفي البساتين أسفل أشجار التفاح أو الحور أو على الجبال أسفل البلوط أو الجوز. عادة ما ينمو الأرقِطيون بالقرب من الجداول وفي مناطق رطبة أخرى. ينمو النبات على ارتفاع 700–1550 م؛ موسم الإزهار والإثمار: يونيو – سبتمبر.

لنبات الأرقِطيون تاريخ طويل من الاستخدام كعشبة طبية؛ حيث تُستخدم الجذور والأوراق والبذور الطازجة والجافة في أغراض طبية. تُستخدم الجذور كمُدرٍّ للبول وكمضادٍ للسموم وكعلاجٍ لالتهاب المفاصل والتهاب الجلد المزمن وكعلاجٍ لحب الشباب والأكزيما والصدفية. استُخدمت البذور في بعض أنواع الحمى والعدوى، مثل النكاف والحصبة.

Artemisia campestris L.

Family Asteraceae

الفصيلة النجمية

field southernwood

شيح *sheeh* ; الأرطميسيا الشجيرية الحقليي

A shrub to 80 cm (31 in) tall, with many stems arising from a woody rootstock, faintly aromatic. Leaves divided into linear segments. Flowers small, 2–3 mm (0.008–0.01 in), borne in capitula c. 4 mm (0.16 in) across, in lax compound inflorescences. Bracts with broad membranous margins.

Field southernwood is native from Europe to China (Gansu) and northwest Iran. In Iraq it is found on undulating depressions in 'haswa' plain, in gravelly and silty soil; alt. 40–160 m (130–525 ft); fl. & fr. Apr., Oct.–Nov.

Field southernwood was known in ancient Mesopotamia as a drug plant called ŠĪḪU(M) (Akkadian). In Iraq, dried flowering branches of this plant have been used to reduce fevers, as an expectorant and vermifuge, as an antiseptic and to heal wounds; it has also been used to induce menstruation, and as an aromatic ointment for the hair.

شجيرة يبلغ طولها 80 سم. متعددة السيقان النامية من جذور خشبية ذات رائحة عطرية خفيفة. الأوراق مُقسَّمة إلى فصوص طولية. الأزهار صغيرة يبلغ طولها 3–2 مم محمولة في رؤيسات يبلغ عرضها 4 مم في نورات مركبة متراخية. القنابات لها حواف عريضة ذات أغشية.

يعود الموطن الأصلي لشجيرة الشيح من أوروبا إلى الصين (مقاطعة قانسو) وشمال غرب إيران. في العراق، تنمو الشجيرة على المنخفضات المتموجة في سهل «الحسوة» في التربة الحصوية والغرينية؛ على ارتفاع 40–160 م؛ موسم الإزهار والإثمار: إبريل وأكتوبر – نوفمبر.

عُرفت شجيرة الشيح في بلاد ما بين النهرين قديمًا كنباتٍ طبيٍّ سُمي ŠĪḪU(M) (في اللغة الأكادية). استُخدمت الفروع المزهرة المُجففة في العراق لتهدئة الحمى، وكطاردٍ للبلغم، وطاردٍ للديدان، ومطهرٍ، ولالتئام الجروح. كما استُخدم النبات لتحفيز الطمث وكمرهمٍ عطريٍّ للشعر.

Asparagus officinalis L.

Family Asparagaceae

garden asparagus

الفصيلة الهليونية

هليون *halyūn* ; الاسبراجوس

An erect herb up to 2 m (6.6. ft) tall, with an underground stem. Foliage feathery, made up of clusters of needle-like cladodes (leaf-like modified stems). Flowers male and female, borne on separate plants, bell-shaped with 6 segments; the male flowers yellow and the female greenish-yellow. Fruit a berry, red, containing up to 6 black seeds.

The garden asparagus is native to Europe, Asia (east to Mongolia) and northwestern Africa (Algeria, Morocco and Tunisia). It is cultivated in Iraq.

Asparagus is grown as an important worldwide crop. Its medicinal properties have been known since antiquity. Extracts of the root have been used for the treatment of urinary and kidney problems, as a diuretic and to treat jaundice and sciatica. The Arabic name *halyūn* is mentioned by Ibn al-Baitar (c. 1240 CE) as a diuretic and as a medicine to remove kidney stones. Asparagus has been used as a treatment for arthritis. It is known as a sedative and as a laxative. Asparagus is also a rich source of glutathione, an antioxidant that is known to boost the immune system, reduce inflammation and maintain the health of the liver.

عشبة منتصبة يبلغ طولها 2 م لها ساق أسفل سطح الأرض. الساق مسطحة ومنتظمة في مجموعات تظهر على شكل أوراق، أما الأوراق الحقيقية فتظهر في شكل حراشف. هناك أزهار تحمل المياسم فقط وأخرى تحمل الأخبية فقط ويكون لها شكل جرسي مكون من 6 فصوص. الأزهار الذكرية صفراء اللون، والأزهار الأنثوية لونها أصفر مائل للخضرة. الثمرة عبارة عن حبة حمراء اللون تضم ست بذورٍ سوداء.

الموطن الأصلي للاسبراجوس هو أوروبا، وآسيا، (شرق منغوليا) وشمال غرب أفريقيا (الجزائر والمغرب وتونس). يُزرع الاسبراجوس في العراق.

يُزرع الاسبراجوس لما له من فائدة معروفة على مستوى العالم؛ فقد عُرفت خصائصه الطبية منذ القدم. استُخدمت مُستخلصات الجذور لعلاج مشكلات المسالك البولية والكلى كمُدرٍّ للبول ولعلاج اليرقان وعرق النسا. ذكر ابن بيطار (نحو عام 1240 م) الاسم العربي (الهليون) للنبات كمُدرٍّ للبول وكعلاجٍ لإزالة حصوات الكلى. يُستخدم الاسبراجوس كعلاجٍ لالتهاب المفاصل ومُهدئٍ ومُلينٍ. يُعرف الهليون أيضًا بكونه مصدرًا غنيًا بالجلوتاثيون، وهو مضاد للأكسدة يعمل على تعزيز جهاز المناعة وتقليل الالتهاب والحفاظ على صحة الكبد.

Astragalus tribuloides Delile

Family Fabaceae

locoweed

الفصيلة البقولية

جرنه *jarnah* ; داتورة عشبية

A small annual herb, branched from the base with prostrate to ascending stems, white, hairy. Leaves pinnate, with 5–10 pairs of oblong to linear leaflets. Inflorescence sessile with flowers borne in groups of 2–6; flowers pea-like, up to 10 mm (0.4 in) long, white to creamy, or white flushed with purple. Pods borne in spreading, star-like clusters, each about 7 mm (0.3 in) long, hairy.

Locoweed is native from southern Russia to Central Asia and India, and North Africa to the Arabian Peninsula. It is found as a common annual in the sub-desert region of Iraq, on bare, rocky desert hills, sandy slopes and plains, gravelly places, appearing after winter rainfall; alt. 50–650 m (160–2,100 ft); fl. & fr. Mar.–May.

In Iraq, the whole plant has been used as an emollient and demulcent.

عشبة صغيرة حولية متفرعة من القاعدة وذات ساق زاحفة إلى صاعدة وزغب أبيض اللون. الأوراق ريشية الشكل لها ٥ ـ ٨ أزواج من الأوراق المستطيلة الطولية. النورات لا عنقية ذات أزهار محمولة في مجموعات مكونة من ٢ ـ ٦ أزهار يشبه شكلها حبة الفول يبلغ طولها ١٠ مم يتدرج لونها من اللون الأبيض إلى اللون الكريمي أو اللون الأبيض المتورد باللون الأرجواني. الأجربة محمولة في تجمعات منبسطة نجمية الشكل مزغبة يبلغ طول كل منها ٧ مم.

الموطن الأصلي للجرنه هو جنوب روسيا إلى وسط آسيا والهند وشمال أفريقيا إلى شبه الجزيرة العربية. الجرنه عشبة حولية شائعة في المنطقة شبه الصحراوية في العراق في التلال الصحراوية الصخرية الجرداء، والمنحدرات والسهول الرملية، والبقاع الحصوية التي تظهر بعد هطول الأمطار شتاءً. تنمو العشبة على ارتفاع 50–650 م؛ موسم الإزهار والإثمار: مارس – مايو.

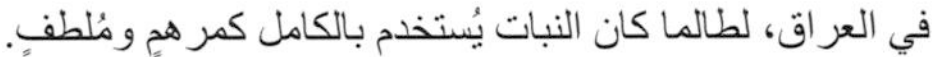

في العراق، لطالما كان النبات يُستخدم بالكامل كمرهمٍ ومُلطفٍ.

Bacopa monnieri (L.) Wettst.

Family Plantaginaceae

الفصيلة الحملية

water hyssop

بربين برى *barbīn barri* ; ماء الزوفا

A perennial herb with 4-angled, smooth, branched, creeping to prostrate stems to 60 cm (24 in) long. Leaves sessile, ovate to spatulate, with entire margins. Flowers 2-lipped, white to blue, borne in lax inflorescences; floral tube 5 mm (0.2 in) long; upper lip 2-lobed; lower lip 3-lobed. Capsule 5 mm (0.2 in) long.

Water hyssop occurs throughout the tropics and subtropics. In Iraq it is found in the western desert area and lower and central alluvial plains, growing in wet and moist locations, near streams, river and stream banks; alt. 40–250 m (130–820 ft); fl. & fr. Apr., Oct.–Nov.

Water hyssop has been used as a remedy for a great number of complaints including various nervous system disorders such as neuralgia, mental illness and epilepsy, as well as for digestive system problems such as indigestion, peptic ulcers, flatulence and constipation. It is also used to treat problems of the respiratory system such as asthma and bronchitis. Furthermore, it has been used to treat infertility. The plant was used in China to deal with impotence, including premature ejaculation and for infertility, gout, pulmonary disease and rheumatism. In some Asian countries, it is used to treat intestinal worms, including filariasis. The plant, boiled to make an infusion, has been used as a laxative and as a diuretic. Fat has been extracted from the herb and this is used locally to relieve joint pain.

Research indicates that this herb is effective in improving mental function, stimulating memory, reducing forgetfulness and restoring mental focus. The herb is considered a neuro-tonic and has been used to treat epilepsy, hysteria, psychosis, Alzheimer's disease, attention deficit hyperactivity disorder (ADHD), memory problems and Parkinson's disease. Owing to its calming effect and non-interference with normal physical activities, it has become a preferred choice for the treatment of ADHD caused by hyperactivity in children. Water hyssop is also considered to be an antioxidant.

عشبة مُعمرة ساقها ناعمة ومتفرعة وزاحفة ذات ٤ زوايا. يبلغ طولها ٦٠ سم. الأوراق لا عنقية، بيضاوية ملعقية الشكل ذات حواف كاملة. الأزهار ذات شفتين ولونها يتراوح من الأبيض إلى الأزرق. الأزهار محمولة في نورات لينة، وأنبوب زهري يبلغ طوله ٥ مم. الشفة العلوية لها فصين، بينما للشفة السفلية ٣ فصوص. علبية البذور يبلغ طولها ٥ مم.

تنمو عشبة ماء الزوفا في المناطق الاستوائية وشبه الاستوائية، وتنمو في العراق في الصحراء الغربية، والسهول الرسوبية المنخفضة والوسطى، وفي البقاع المطيرة والرطبة إلى جوار الجداول والأنهار وضفافها. ينمو النبات على ارتفاع 250–40 م؛ موسم الإزهار والإثمار: إبريل وأكتوبر - نوفمبر.

تُستخدم عشبة ماء الزوفا كعلاجٍ لعددٍ كبيرٍ من الشكاوى؛ بما في ذلك اضطرابات الجهاز العصبيِ المختلفة مثل الألم العصبي والأمراض العقلية والصرع، وكذلك اعتلالات الجهاز الهضمي مثل عسر الهضم والقرحات الهضمية وانتفاخ البطن والإمساك، كما أن العشبة مفيدة لمشكلات الجهاز التنفسي مثل الربو والتهاب الشعب الهوائية. إضافة إلى ذلك، استُخدم النبات لعلاج العقم، وقد استُخدم في الصين في التعامل مع حالات العجز الجنسي؛ بما في ذلك القذف المبكر والعقم والنقرس وأمراض الرئة والروماتيزم. يُستخدم النبات، في بعض البلدان الآسيوية لعلاج الديدان المعوية؛ بما في ذلك داء الخيطيات. ويُستخدم مغلى النبات لتحضير منقوع يُستخدم كمُلينٍ ومُدرٍ للبول. تُستخلص الدهون من العشبة وتُستخدم محليًا لتخفيف آلام المفاصل.

تشير الأبحاث إلى أن هذه العشبة فعَّالة في تحسين الوظيفة العقلية وتنشيط الذاكرة والحدّ من النسيان واستعادة التركيز الذهني. تعتبر العشبة مُنشطًا عصبيًا وتُستخدم لعلاج الصرع والهستيريا والذهان ومرض الزهايمر ونقص الانتباه ومشكلات الذاكرة وداء باركنسون (الشلل الرعاش). ونظرًا لأثر العشبة المهدئ وعدم تدخلها في الأنشطة البدنية العادية للفرد، أصبحت الخيار الأفضل لعلاج نقص الانتباه الناجم عن فرط النشاط عند الأطفال، كما تُعد عشبة ماء الزوفا من مضادات الأكسدة.

Bellis perennis L.

Family Asteraceae

الفصيلة النجمية

daisy

zahar al lulu زهر الولو ; الأقحوان ;
zahr al rabee' زهر الربيع

A perennial scapose (with stems growing direct from the ground) herb, up to 25 cm (10 in), mat-forming, with a root system of many fibrous rootlets. Leaves spatulate, tapering below to a winged petiole, margins entire to shallowly serrate or dentate, some leaves usually toothed. Bracts (phyllaries) dark green, usually with an apical tuft of white hairs. Ray florets white or pink-tinged beneath. Achenes ± 2 mm (0.008 in).

The daisy is found throughout Europe, Cyprus, Turkey, Israel and Palestine, Iraq, Transcaucasia and Macaronesia (Azores, Madeira); it is naturalised and weedy in Pakistan, India, Australia, New Zealand and North and South America. In Iraq the daisy is widespread in mountain regions, but rare in the foothills and plains. It grows in damp places on hillsides and in valleys, damp grassy places and orchards, irrigated gardens and by or even in streams, usually on clayey soil; alt. 800–1,370 m (2,600–4,500 ft); fl. Mar.–June.

Leaves and flowers of the daisy have been used as a treatment for chronic colds and as a tonic for the digestive system. The flowers have been used to treat chest problems and in healing wounds, and the leaves to stop bleeding and heal wounds. The leaves have also been also used as a diuretic, blood purifier and anti-colic. The plant has also been used for the treatment of laryngitis, sore throat and bronchitis; it is recorded to be beneficial for the treatment of sprains, abscesses, high blood pressure, jaundice and kidney diseases.

عشبة مُعمرة ذات ساق عارية بلا أوراق تنمو مباشرة من الأرض ويبلغ طولها 25 سم. تنمو العشبة في شكل بساط، ولها نظام جذري يتألف من الكثير من الجذور الليفية. الأوراق ملقية الشكل، ومستدقة من أسفل ذات سويقات مجنحة والحواف بالكامل مشرشرة أو مسننة قصيرة. عادة ما تكون بعض الأوراق مسننة. لون القنابات (الأوراق النباتية) أخضر داكن. عادة توجد شوشة قمية من الشعر الأبيض. زهيرات بيضاء أو زهرية اللون من جهتها السفلية. يبلغ طول الثمرة ±2 مم.

يُزرع الأقحوان في جميع أنحاء أوروبا وقبرص وتركيا وفلسطين والعراق والقوقاز وماكارونيسيا (جزر الأزور وماديرا)؛ وهو عشبة مستوطنة في باكستان والهند وأستراليا ونيوزيلندا وأمريكا الشمالية والجنوبية. ينتشر الأقحوان في العراق في المناطق الجبلية، ويندر في التلال والسهول، وينمو في الأماكن الرطبة على سفوح التلال والوديان والأماكن العشبية الرطبة والبساتين والحدائق المروية وإلى جوار الجداول أو في مجرى الجداول ذاتها. عادة ما يحتاج النبات إلى تربة طينية. ينمو الأقحوان على ارتفاع 800–1370 م؛ موسم الإزهار: مارس – يونيو.

تُستخدم الأوراق والأزهار كعلاجٍ لنزلات البرد المزمنة وكمُنشطٍ للجهاز الهضمي. تُستخدمِ الأزهار في علاج أمراض الصدر ولالتئام الجروح، بينما تُستخدم الأوراق لوقف النزيف والتئام الجروح. استُخدمت الأوراق أيضًا كمُدرٍّ للبول وتنقية الدم وكمضادٍ للمغص، كما استُخدم النبات لعلاج التهاب الحنجرة والتهاب الحلق والتهاب الشعب الهوائية. سُجِّل النبات كعشبة مفيدة في علاج الالتواءات والخراجات وارتفاع ضغط الدم واليرقان وأمراض الكلى.

Bidens tripartita L.

Family Asteraceae

water agrimony

الفصيلة النجمية

القنب المائي; *qinnab maiy* قنب ماىٔ;
tel maiy تيل ماىٔ

An erect, branched, annual herb up to 80 cm (31 in); stems hairy. Leaves lanceolate-ovate in outline, 2(–5)-lobed, lobes narrowly ovate with dentate margins. Inflorescence comprising solitary capitula, yellowish-brown, terminal and in the axils of stems subtended by about 6 leafy bracts. Achenes with backward-pointing stiffly haired awns.

Water agrimony is native to the temperate Northern Hemisphere. In Iraq, it is found in the lower forest and grassland zone, usually near water by streams, ditches and riverbanks, and in irrigated plantations; alt. 50–1,150 m (160–3,800 ft); fl. & fr. June–Sept.

Water agrimony has been recorded as a medicinal herb in Iraq but its use is infrequent. It has been used for the treatment of bladder and kidney problems as it is a diuretic, and also to stop uterine bleeding. Water agrimony is good for the treatment of digestive problems, especially gastric ulcer, diarrhoea and ulcerative colitis. Mixed with other herbs, such as ginger, it can be used to relieve flatulence.

عشبة حولية منتصبة متفرّعة، يبلغ طولها 80 سم؛ الساق مزغبة. الأوراق بيضاوية رمحية الشكل مُكوَّنة من (5–)2 فصوص. الفصوص بيضاوية رفيعة ذات حواف مسننة. النورات تتكون من رؤيسات منفردة أطرفها بنية مائلة إلى الصفرة. يقابل محاور السوق حوالي 6 قنابات ورقية. الثمرة ذات حسك مزغب متصلب متجه نحو الخلف.

الموطن الأصلي للقنب المائي هو نصف الكرة الشمالي المعتدل. في العراق، يوجد النبات في منطقة الغابات المنخفضة والمراعي، وعادة ما ينمو بالقرب من الجداول والقنوات وضفاف الأنهار وفي المزارع المروية. ينمو النبات على ارتفاع 50–1150 م؛ موسم الإزهار والإثمار: يونيو– سبتمبر.

سُجّل القنب المائي كعشبة طبية في العراق، ولكنه ليس كثير الاستخدام. استُخدم النبات كعلاجٍ لمشكلات المثانة والكلى وكمُدرٍّ للبول، كما استَخدم لوقف نزيف الرحم. يُعرف النبات بفائدته في علاج مشكلات الهضم، لا سيّما القرحات الهضمية والإسهال والتهاب القولون القرحي. يُخلط القنب المائي مع أعشاب أخرى مثل الزنجبيل، ويُستخدم المزيج لعلاج انتفاخ البطن.

Brassica nigra (L.) W.D.J.Koch

Family Brassicaceae

black mustard

الفصيلة الكرنبية

الخردل الأسود *khardii*

An erect annual herb to 1 m (3.3 ft) tall. Lower leaves deeply divided to a terminal lobe and several lateral lobes, margins dentate; upper leaves becoming much smaller. Flowers bright yellow, in terminal inflorescences. Fruit a narrow pod, 10–20 mm (0.4–0.8 in) long and about 2 mm (0.008 in) broad. Seeds minute, about 1 mm in diameter, globose, dark brown to blackish.

Black mustard is native from Western Europe east to China and from the Mediterranean Basin south to Ethiopia; it has been introduced to and cultivated in many other regions. In Iraq, black mustard is found in the forest and lower hills, on mountain slopes, in oak forest, and as a weed in fields, plantations and waste places; alt. 50–1,700 m (160–5,600 ft); fl. & fr. Mar.–June.

Black mustard has been cultivated for oil and for its edible vegetative parts since Classical times.

Carbonised seeds of black or white mustard dating back to 5000 BCE have been recovered from Iraq. It is called KASÛ (Akkadian) and defined as a food plant in ancient Mesopotamia. Mustard was mentioned by Dioscorides for the medicinal value of its oil (1–47 Sinapelaion) and for the preparation of Sinapinum, which is '... good for diseases of long duration drawing out faulty fluids from within'.

Black mustard has been used extensively and was the main source of mustard, but more recently this species has been replaced by high-yielding cultivars of brown- or yellow-seeded species (*Brassica juncea, B. hirta*). Mustard is commonly used in southwest Asian cuisines, and mustard oil is used as cooking oil. The shoots and leaves are used as a vegetable, and the plant is highly regarded in southwest Asia, where the seeds and oil are used to treat various ailments from coughs and colds to cardiovascular diseases.

عشبة حولية منتصبة يبلغ طولها 1 م. الأوراق السفلية مُقسمة تقسيمًا عميقًا وذات فص طرفي وعدة فصوص جانبية. حواف الأوراق مسننة. الأوراق العلوية أصغر بكثير. الأزهار صفراء فاقعة اللون في نورات طرفية. الثمرة عبارة عن جراب ضيق يبلغ طوله 10–20 مم وعرضه ٢ مم. البذور دقيقة وكروية الشكل ولونها بني داكن إلى مائل إلى السواد ويبلغ قطرها نحو 1 مم.

الموطن الأصلي للخردل الأسود هو أوروبا الغربية شرقًا إلى الصين ومن حوض البحر الأبيض المتوسط جنوبًا إلى إثيوبيا. وقد استُحدثت زراعة الخردل الأسود في مناطق أخرى كثيرة. في العراق، يوجد الخردل الأسود في الغابة والتلال المنخفضة، وعلى المنحدرات الجبلية، وفي غابات البلوط، كما ينمو كحشيشة في الحقول والمزارع والأراضي المهملة. ينمو الخردل الأسود على ارتفاع 50–1700 م؛ موسم الإزهار والإثمار: مارس – يونيو.

يُزرع الخردل الأسود للاستفادة بزيته والأجزاء الخضرية الصالحة للأكل منذ القدم.

أُخذت البذور المُكربنة للخردل الأسود أو الأبيض من العراق منذ عام 5000 ق.م. وقد سُمي الخردل الأسود في اللغة الأكادية باسم (KASÛ) وعُرف كنباتٍ صالحٍ للأكل قديمًا في بلاد ما بين النهرين. ذكر ديسقوريدوس الخردل لما لزيته من قيمة طبية (1–47 سينابلايون) وتحضير Sinapinum الذي « .... يصلح كعلاجٍ للأمراض طويلة الأجل ويسحب السوائل الضارة من الداخل».

كان الخردل الأسود يُستخدم على نطاقٍ واسعٍ؛ حيث كان يُعد المصدر الرئيسي للخردل؛ بيْد أنه استُبدل في الآونة الأخيرة بالأصناف عالية الغلة للأنواع ذات البذور بنية اللون أو صفراء اللون (*Brassica juncea, B. hirta*). ويشيع استخدام الخردل في مطابخ جنوب غرب آسيا، ويُستخدم زيت الخردل في طهي الطعام. تُستخدم البراعم والأوراق كخضراوات، ويحظى النبات بتقديرٍ بالغٍ في جنوب غرب آسيا. تُستخدم بذور الخردل والزيت المُستخلص منه لعلاج أمراض مختلفة من السعال ونزلات البرد إلى أمراض القلب والأوعية الدموية في جنوب غرب آسيا.

In Iraq, the seeds of black mustard have been used as a heart/lung tonic. The seeds are also used for headache and dizziness and as an anti-neuralgic agent. When used as a gargle, an infusion of seeds has been found to ease coughing and tonsillitis; infusions have also been used to treat scurvy and gout and as a laxative. The seeds have been reported to expedite recovery from measles. Mustard oil has been used to treat hardening of the arteries and hypertension, and as a poultice applied on the body for rheumatism and ligament pain.

في العراق، استُخدمت بذور الخردل الأسود ككمادة منعشة للقلب/ الرئة، واستُخدمت البذور للصداع والدوار وكعاملٍ مُضادٍ للألم. يُستخدم منقوع بذور الخردل الأسود كغرغرة لتهدئة السعال والتهاب اللوزتين. كما استُخدم المنقوع أيضًا لعلاج الاسقربوط والنقرس وكمُلينٍ. عُرفت بذور الخردل الأسود بدورها في تسريع التعافي من الحصبة. واستُخدم زيت الخردل في علاج تصلب الشرايين وارتفاع ضغط الدم، وككمّادٍ يوضع على الجسم لعلاج الروماتيزم وآلام الأربطة.

Bryonia multiflora Boiss. & Heldr

Family Cucurbitaceae

الفصيلة القرعية

red bryony

مرزو *marazho* (Kurd.) ; بريوني أحمر

A climber to several metres high with a tuberous root and glabrous to pubescent stems, scrambling over trees and shrubs. Leaves with simple tendrils, lamina deeply palmately 5-lobed with lobes sometimes twice-lobed, with margins irregularly serrate, punctate or pubescent below. Flowers yellowish-green; male and female flowers on separate plants. Male flowers in elongated racemes; female flowers in shorter, dense racemes. Fruit a berry, globose, 7–8 mm (0.27–0.3 in) in diameter, red when ripe, shiny.

Bryony is native to Turkey, southern Syria, northern Iraq, west and southwest Iran. In Iraq it is found on rocky slopes in the mountainous north; alt. 450–1,400 m (1,500–4,600 ft); fl. & fr. Apr.–June.

Red and white bryony are mentioned in the earliest Arabic pharmacopoeia known to have survived, *al Aqrābādhin al-saghīr* by Sābūr ibn Sahl (a Nestorian physician and pharmacologist, d. 869 CE), indicating that bryony was used by medical practitioners of those times. The main uses listed are as a diuretic, as a laxative and for cleansing. The tuberous roots have also been used to treat diabetes. Chakravarty & Jeffery (2008) note that there is no record of the use of this plant, but it is likely to have been used in earlier times as a medicinal herb. Al-Rawi & Chakravarty (1964) (p. 20) lists *Bryonia dioica* (white bryony of Europe) in *Medicinal Plants of Iraq*, but this species is not recorded in Iraq. Al-Douri (2000) also lists *B. dioica* as a medicinal herb, but it is likely to be *B. multiflora*.

نبات متسلق يبلغ طوله عدة أمتار. النبات ذو جذر درني والساق ملساء أو مغطاة بشعيرات خفيفة، ومتسلقة ومعترشة أعلى الأشجار والشجيرات. الأوراق ذات معاليق بسيطة، ورقائق راحية عميقة مكونة من 5 فصوص، وأحيانًا ما تكون مزدوجة الفصوص، وذات حواف غير منتظمة مسننة أو مثقبة أو مزغبة من جهتها السفلية. الأزهار خضراء اللون مائلة للصفرة. هناك نباتات تحمل الأعضاء التناسلية الذكرية وأخرى تحمل الأعضاء التناسلية الأنثوية. أزهار النباتات التي تحمل الأعضاء التناسلية الذكرية طويلة، بينما الأزهار الأنثوية أقصر طولًا وكثيفة. الثمرة تشبه التوت؛ كروية الشكل قطرها 7-8 مم، يصبح لون الثمرة أحمر قاني عند نضجها.

الموطن الأصلي للبريوني الأحمر هو تركيا وجنوب سوريا وشمال العراق وغرب وجنوب غرب إيران. في العراق، ينمو النبات على المنحدرات الصخرية في المناطق الجبلية من شمال العراق. ينمو النبات على ارتفاع 1400–450 م؛ موسم الإزهار والإثمار: إبريل - يونيو.

ذكر سابور بن سهل (طبيب وصيدلي نسطوري - توفي عام 869 م) البريوني الأحمر والأبيض في الأقرباذين الصغير، أقدم دستور عربي للصيدلة. وأشار الكتاب إلى أن النبات كان يُستخدم بواسطة ممارسي الطب في هذه الآونة. وقد جاءت الاستخدامات الرئيسية المُدرجة كالتالي: مُدرّ للبول ومُلين ومُطهر. استُخدمت الجذور الدرنية لعلاج داء السكري. أشار شاكرافارتي وجيفري في كتاب نباتات العراق (*Flora of Iraq*) 4(1): الصفحة 200، إلى أنه لا يوجد سجّلٌ لاستخدام هذا النبا ت، غير أنهما رجّحا احتمال استخدامه في أوقات سابقة كعشبٍ طبيٍّ. أدرج راوي (الصفحة 20) نبات *Bryonia dioica* (البريوني الأبيض في أوروبا) ضمن النباتات الطبية في العراق - لكن هذا النوع غير مُسجَّلٍ في العراق. ذكر الدوري (2001) أيضًا البريوني الأحمر والأبيض (*B. dioica*) كعشبٍ طبيٍّ، ولكن من المُحتمل أن يكون نبات متعدد الأزهار (*B. multiflora*).

Calotropis procera (Aiton) W.T.Aiton

Family Apocynaceae

الفصيلة الدفلية

mudar plant

الد يباج *al dibaj*; اشكر *ashkar*;

A branched, robust, grey-green, shrub, up to 2 m (6.6 ft) tall, somewhat succulent, the whole plant often appearing farinose; woody base covered with a corky bark; plant exudes copious milky sap when snapped. Leaves opposite, ovate or broadly obovate, glaucous. Flowers several, borne in umbellate cymes; flowers with a campanulate corolla, lobes white with purple tips; corona-lobes with a basal spur, white or purple, cleft radially in the upper half. Fruit a follicle, 80 mm–13 cm (3–5 in) long, ovoid to subglobose, with one side somewhat flattened. Seeds with an apical tuft of silky hairs, dispersed by wind.

The mudar plant is native to North and tropical Africa to Indo-China (naturalised in arid areas elsewhere). It is sometimes cultivated in Iraq, but is also sub-spontaneous in the drier parts, especially in sandy and gravelly locations; alt. 100–500 m (330–1,640 ft); fl. & fr. Nov.

An infusion of the root bark of the mudar plant has been used as antispasmodic, as a sudorific and as a tonic; in large doses it has been used as an emetic. The flowers have been used as a digestive and as a tonic. In the Arabian Peninsula, the leaves and latex have been used for treating wounds, extracting pus from wounds, treating scorpion stings and to relieve pain and swelling. To strengthen muscles affected by paralysis, leaves of *ashkar* mixed with clove oil, the fruit of *Terminalia catappa* and seeds of *Nigella sativa* are heated and rubbed over paralysed limbs.

شجيرة متفرعة وقوية ولونها رمادي مائل إلى الخضرة. يبلغ طولها مترين. نبات كثيف الأوراق إلى حدٍ ما. غالبًا ما يظهر النبات بأكمله في شكل دقيقي (نشوي). للنبات قاعدة خشبية مغطاة بلحاء من الفلين. النبتة تفرز عصارة حليبية غزيرة عند قطعها. الأوراق معكوسة وبيضاوية أو كثيرًا ما تكون بيضاوية مقلوبة، ولونها أخضر باهت. الأزهار محمولة في نورات خيمية تحتوي على عدة أزهار. التويج ناقوسي الشكل له فصوص بيضاء ذات أطراف أرجوانية. الفصوص التاجية ذات نتوء قاعدي ولونها أبيض أو أرجواني، مشقوقة قطريًا في النصف العلوي. الثمرة عبارة عن جريب طوله 80 مم إلى 13 سم، وبيضاوي الشكل إلى شبه كروي وله جانب واحد مسطح إلى حدٍ ما. البذور ذات خصلة قمية حريرية تتطاير بفعل الرياح.

الموطن الأصلي للديباج هو شمال أفريقيا وأفريقيا الاستوائية إلى الهند الصينية (يستوطن النبات المناطق القاحلة في أماكن أخرى). يُزرع أحيانًا في العراق، ولكنه أيضًا ينمو على نحو شبه تلقائي في المناطق الجافة، لا سيّما في البقاع الرملية والحصوية. ينمو النبات على ارتفاع 100–500 م؛ موسم الإزهار والإثمار: نوفمبر.

يُستخدم منقوع لحاء جذر النبات كمُضادٍ للتشنج، ومُعرِّق، ومُقوٍّ، كما يُستخدم بجرعات كبيرة كمُقَيِّئ. تُستخدم الأزهار كمُهضمٍ ومُنشطٍ. في شبه الجزيرة العربية، تُستخدم الأوراق وعصارة النبات في شبه الجزيرة العربية لعلاج الجروح واستخراج الصديد من الجروح وعلاج لدغات العقارب وتسكين الألم والتورُّم. لتقوية العضلات المصابة بالشلل، تخلط أوراق الديباج بزيت القرنفل وثمرة تيرميناليا كاتابا وبذور حبة البركة ويُسخن الخليط ويُفرك على الأطراف المصابة.

Chrozophora tinctoria (L.) A.Juss

Family Euphorbiaceae

tournsole, katsol, folium

الفصيلة الفربيونية

zurraij زريج; *nīl* نيل

An annual herb, often woody at the base, 30 cm–1 m (12–39 in) tall; the whole plant stellate-hairy. Leaves ovate to ovate-lanceolate with repand (wavy) -denticulate margins. Inflorescence terminal; male and female flowers separate, small, on the same inflorescence. Male flowers with filaments united into a column; female flowers with ovary densely covered with peltate (shield-shaped) scales. Capsule about 8 mm (0.3 in) in diameter, warty and covered with peltate scales.

Tournsole is native from North Africa and the Mediterranean Basin to northwest India. In Iraq it is common in the lower forest zone and the desert regions, found on rocky slopes, in oak forests, on disturbed and waste ground in the plains, in plantations, orchards, by roadsides in ditches, on clay, loamy and sandy soils; alt. up to 1,000 m (3,300 ft); fl. & fr. Apr.–Aug. & July–Nov.

Tournsole is recorded to be cathartic and emetic. In Iraq, decoctions of the whole plant have been used as a laxative and emetic. The plant (fruit) yields a dark blue dye known as turnsole, katsol or folium blue, which was used in medieval illuminated manuscripts for the blue and purple colours. The extracted dye has been said to be stored with cloth and dried as a watercolour; when needed, a piece of cloth was then cut and the paint extracted with water. The dye has also been used in the production of Dutch cheese.

عشبة حولية، غالبًا ما تكون خشبية عند القاعدة. طول العشبة ٣٠ سم - ١ متر. العشبة بالكامل مغطاة بشعيرات نجمية. الأوراق بيضاوية إلى بيضاوية - رمحية ذات حواف متموجة - مُسننة. النورات طرفية، ويوجد في نفس النورة الأزهار التي تحمل الأعضاء الذكرية والأزهار التي تحمل الأعضاء الأنثوية على نحوٍ منفصلٍ. النورات صغيرة الحجم. الأزهار التي تحمل الأعضاء الذكرية متصلة بخيوط في عمود، بينما الأزهار التي تحمل الأعضاء الأنثوية ذات مبيض مغطى بقشور ترسية الشكل كثيفة. علبة البذور قطرها حوالي 8 مم، ثؤلولية ومغطاة بقشور ترسية.

الموطن الأصلي للنبات هو شمال أفريقيا وحوض البحر الأبيض المتوسط إلى شمال غرب الهند. في العراق، ينتشر النبات في منطقة الغابات المنخفضة والمناطق الصحراوية، وينمو أيضًا على المنحدرات الصخرية في غابات البلوط، على التربة المضطربة والأراضي المهملة في السهول، وفي المزارع والبساتين، وعلى جانبي الطرق في الخنادق. ينمو النبات في تربة طينية وطميية ورملية على ارتفاع 1000 م؛ موسم الإزهار والإثمار: إبريل – أغسطس ويوليو- نوفمبر.

سُجِّل النبات كمُسهلٍ ومُقَيِّئ. في العراق، يُستخدم مُستخلص النبات بأكمله بالإغلاء كمُلينٍ ومُقَيِّئ. يُنتج النبات (الثمرة) صبغة زرقاء داكنة تُعرف باسم turnsole، أو katsol، أو folium blue. استُخدمت هذه الصبغة في العصور الوسطى في المخطوطات المضيئة للونين الأزرق والبنفسجي. أُشير إلى أنه ينبغي للصبغة المستخرجة أن تُخزَّن مع قطعة قماش وأن تُجفف كألوان مائية. عند الحاجة، تُقطَّع قطعة من القماش ويُستخلص الطلاء بالماء. استُخدمت الصبغة أيضًا في إنتاج الجبن الهولندي.

Cistanche tubulosa (Schrenk.) Wight

Family Orobanchaceae

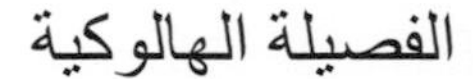

الفصيلة الهالوكية

desert hyacinth

thanoon ثنون

A parasitic herb with an unbranched, succulent stem to 70 cm (28 in) tall; sometimes clumped. Leaves reduced to scale-like bracts, grey below, purplish above, overlapping at base. Flowers bright yellow, often purplish in bud, borne in terminal spikes; corolla tubular, 2-lipped, the upper lip 2-lobed, the lower 3-lobed, lobes spreading. Anthers with white-woolly hairs.

Native to the Middle East (full distribution is unclear owing to confusion with other similar species). Found in the desert regions in Iraq, on sandy and saline soils, and coastal areas, parasitic on various salt bushes (especially *Haloxylon salicornicum* (=*Hamada salicornica*), and apparently also *Capparis spinosa* and *Zygophyllum coccineum*); flowering after winter and spring rainfall; alt. up to 100 m (330 ft); fl. & fr. Apr.–June; Sept.–Nov.

The fleshy stem boiled in water has been used (eaten) to treat diarrhoea. In Baluchistan (southwest Pakistan), the plant was used to treat diarrhoea and applied on skin to treat sores.

عشب طفيلي، ذو ساق عصارية غير متفرّعة. يبلغ طوله 70 سم وفي بعض الأحيان يكون متكتلًا. الأوراق على شكل قنابات تشبه القشور، رمادية اللون من الجهة السفلية، وأرجوانية من الجهة العلوية، ومتداخلة من عند القاعدة. الأزهار صفراء فاقعة اللون، وغالبًا ما تكون أرجوانية من عند البراعم، ومحمولة في سنابل طرفية. التويج أنبوبي الشكل مكون من شفتين. الشفة العلوية مكونة من فصين والسفلية من ٣ فصوص، والفصوص منبسطة. المتك ذو شعر أبيض اللون صوفي الملمس.

الموطن الأصلي للعشب هو الشرق الأوسط (التوزيع الكامل للعشب غير واضح بسبب الخلط بينه وبين أنواع أخرى متشابهة). ينمو العشب في المناطق الصحراوية في العراق، في التربة الرملية والمالحة، والمناطق الساحلية. يتطفل العشب على أجام (*Hamada salicornica*)=*Haloxylon salicornicum* ملحية متنوعة (لا سيّما ، وعلى ما يبدو يتطفل أيضًا على *Capparis spinosa* ، و *Zygophyllum* coccineum). موسم الإزهار بعد هطول الأمطار في الشتاء والربيع. ينمو النبات على ارتفاع ١٠٠ متر؛ موسم الإزهار: إبريل–يونيو، سبتمبر - نوفمبر.

يُستخدم الساق اللحمي المغلي في الماء (يؤكل) لعلاج الإسهال. في بلوشستان (جنوب غرب باكستان)، يُستخدم النبات لعلاج الإسهال ويُدهن على الجلد لعلاج القروح.

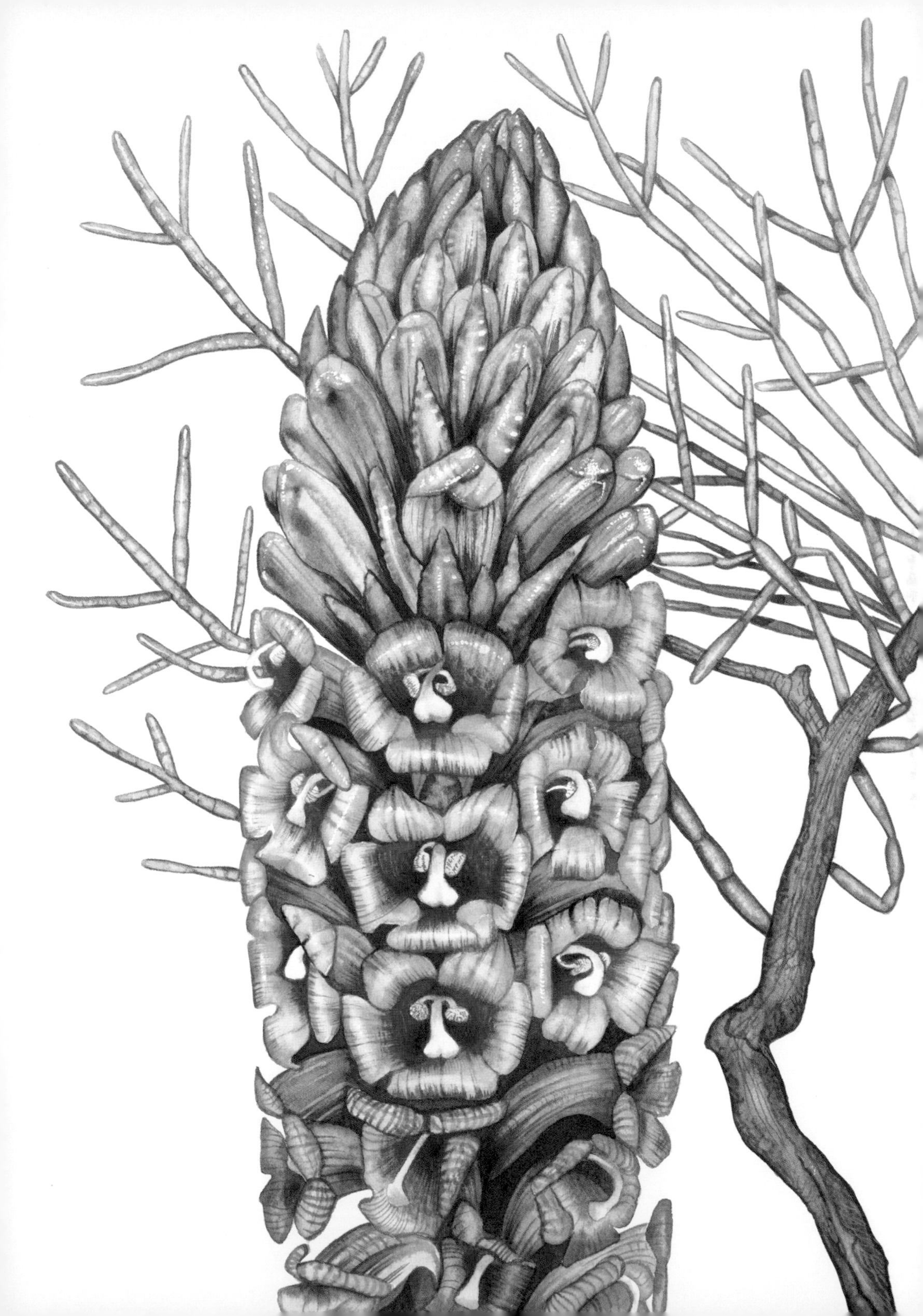

Citrullus colocynthis (L.) Schrad.

Family Cucurbitaceae

الفصيلة القرعية

colocynth, bitter apple, bitter gourd

handhal حنظل; *hadaj* حدج;
marārat as-sahāri مراة الصحاري;
shiri شري

A perennial herb with spreading, prostrate, creeping stems to 1 m (3.3 ft) with tendrils; whole plant scabrid (rough-textured). Leaves alternate, 3–5-lobed. Flowers unisexual, yellow, solitary, about 20 mm (0.8 in) across. Fruit a globose gourd, 50–90 mm (2–3.5 in) across, green and yellow striped and marbled when immature, ripening yellow, smooth. Seeds imbedded in white bitter flesh, eventually the fruit becoming partially hollow and seeds loose and rattling inside.

Colocynth is native to Macaronesia and the Mediterranean Basin east to Myanmar, and from northeast tropical Africa to northern Kenya. In Iraq it is frequent in the desert regions, on sandy and gypsophilous substrates, and on sandy and gravelly plains, often in wadis or depressions; alt. up to 400 m (1,300 ft); fl. Mar./Apr.–Oct./Nov, fr. May/June–Dec.–Jan.

The medicinal properties of colocynth have been known since early times. Chakravarty & Jeffery (1980, p. 194) note that Arabic names used for the plant mentioned by Ibn al-Baitar (c. 1240 CE) include *bushbush* بشبش as the leaf of the plant, *hadaj* حدج as the unripe gourd, *kabast* كبست as the pulp and *habad* حبد as the seeds. The pulp of the fruit has violent purgative properties and has been considered useful in constipation and fever, and for treating intestinal parasites. The roots have been used for jaundice, urinary problems and rheumatism.

Colocynth is well known and has been used medicinally in the Arabian Peninsula since early times. In Yemen, parts of the fruit are worn in amulets to protect against 'evil eye' and a whole fruit placed in a new house for the same purpose.

عشب مُعمر ذو ساق زاحفة يبلغ طولها متر واحد. الساق ذات معاليق، والنبات بأكمله خشن الملمس. الأوراق متناوبة لها 3–5 فصوص. الأزهار أحادية الجنس، وصفراء اللون، ومفردة، وعرضها حوالي 20 مم. الثمرة قرعية كروية، عرضها 90–50 مم، مخططة باللونين الأخضر والأصفر، ورخامية قبل نضجها. تصبح الثمرة صفراء اللون ناعمة الملمس عند نضجها. البذور مدمجة في لب الثمرة ذي اللون الأبيض، والمذاق المُرّ. في النهاية، تصبح الثمرة مجوفة جزئيًا والبذور مفككة مخشخشة من الداخل.

الموطن الأصلي للنبات هو ماكارونيسيا وحوض البحر الأبيض المتوسط شرقًا إلى ميانمار، ومن شمال شرق أفريقيا الاستوائية إلى شمال كينيا. في العراق، ينتشر النبات في المناطق الصحراوية، على الطبقات الرملية والجصية، والسهول الرملية والحصوية، غالبًا في الوديان أو المنخفضات. ينمو النبات على ارتفاع 400 م؛ موسم الإزهار: مارس / إبريل - أكتوبر / نوفمبر، موسم الإثمار: مايو / يونيو – ديسمبر / يناير.

عُرفت الخصائص الطبية للنبات منذ القدم. وقد أشار شاكرافارتي وجيفري في كتاب نباتات العراق (*Flora of Iraq*) (4(1): الصفحة) أن الأسماء العربية المُستخدمة للنبات قد ذكرها ابن البيطار (نحو عام 1240 م) بما في ذلك شجيرة بشبش (*bushbush*) في إشارة إلى ورقة النبات، وحدج (*hadaj*) في إشارة إلى القرع غير الناضج، وكبست (*kabast*) في إشارة إلى اللب وحبد (*habad*) في إشارة إلى البذور. لب الثمرة له خصائص مُسهلة قوية وكان ذا جدوى في حالات الإمساك والحمى وعلاج الطفيليات المعوية. استخُدمت الجذور لعلاج اليرقان واعتلالات المسالك البولية والروماتيزم.

يُعرف النبات جيدًا ويُستخدم طبيًا في شبه الجزيرة العربية منذ العصور القديمة. في اليمن، توضع أجزاء من الثمار في تمائم تُرتدى للحماية من «العين الشريرة» (درءًا للحسد) وتوضع ثمرة كاملة في البيوت الجديدة للحماية من الحسد.

Clinopodium graveolens (M.Bieb.) Kuntze

Family Lamiaceae

الفصيلة الشفوية

wild basil

reihan barri ريحان بري

An annual herb with erect to spreading stems, 8–25 cm (3–10 in), glandular. Leaves ovate-orbicular, with shallow serrate margins. Inflorescence of 4–8 flowered whorls in spikes. Calyx tubular, swollen at base, 2-lipped, upper lip 3-toothed, the lower 2-toothed. Corolla 2-lipped, pink. Nutlets oblong, glabrous.

Wild basil is native from the eastern Mediterranean to Central Asia. In Iraq, it is common in the western and northwestern parts of the forest zone, on limestone in oak forest, on rocky mountains and on mountain slopes; alt. 600–1,440 m (2,000–4,700 ft); fl. Apr.–May & June.

Infusions of the flowers of the wild basil have been used to treat flatulence, colonitis and stomach pains, and as a purgative. The seeds have been used as an aphrodisiac and to revive consciousness.

عشبة حولية. الساق منتصبة إلى زاحفة غدية. يبلغ طول الساق ٨-٢٥ سم. الأوراق بيضاوية دائرية، وحوافها مسننة ضحلة. النورات تحتوي على ٤-٨ جدلات مزهرة في سنابل. كأس الزهرة أنبوبي الشكل، ومنتفخ من عند القاعدة، وذو شفتين؛ الشفة العلوية لها ٣ أسنان، والشفة السفلية لها زوج من الأسنان. التويج وردي اللون وذو شفتين. الجوزيات مستطيلة الشكل وملساء الملمس.

الموطن الأصلي للريحان البري هو شرق البحر الأبيض المتوسط إلى آسيا الوسطى. في العراق، ينتشر هذا النبات في المناطق الغربية والشمال الغربية من منطقة الغابات، وينمو أيضًا على الحجر الجيري في غابات البلوط، وعلى الجبال الصخرية والمنحدرات الجبلية. ينمو النبات على ارتفاع 600–1440 م؛ موسم الإزهار: إبريل - مايو ويونيو.

استُخدم منقوع أزهار الريحان البري لعلاج انتفاخ البطن والتهاب القولون وآلام المعدة وكمُلينٍ. وقد استُخدمت البذور كمُنشطٍ جنسيٍّ ولاستعادة الوعي.

Crocus sativus L.

Family Iridaceae

الفصيلة السوسنية

saffron

za'farān زعفران

A small, bulbous perennial with the corm covered in closely netted fibres. Leaves 9–11, greyish-green, glabrous. Flowers purple, perianth segments darker towards the base. Style exceeding the perianth, orange-red, divided into 3 red arms, expanded at the apex, aromatic.

Saffron is not known in the wild; the cultigen originates from Greece. Saffron has been introduced and is widely cultivated across Europe, Morocco, Turkey, Iran, Pakistan and the western Himalayas. It is reported to have been cultivated in Iraq; Mathew in *Flora of Iraq* 8 (1985, p. 2121) notes that the place name Za'faraniyah may refer to a place where it was once cultivated.

The orange tops of the styles and stigmas are the parts of the plant used for medicinal and other purposes. Saffron has been widely used medicinally since Greek and Roman times. Currently, the principal use of saffron is as a flavouring agent used in cooking, especially in Iranian cuisine, and as a dye. An infusion of saffron has been used to reduce fevers and headaches, to treat melancholia, as a sedative, and in the treatment of an enlarged liver.

نبات مُعمر صغير بصلي ذو قرم مغطى بألياف متشابكة ومتقاربة. للنبات 11–9 ورقة ملساء خضراء مائلة للون الرمادي. الأزهار أرجوانية اللون. الغلاف الزهري أكثر دكنة باتجاه القاعدة. المحيط الذي يتجاوز الغلاف الزهري، برتقالي – أحمر اللون، ومنقسم إلى 3 أذرع حمراء اللون، وممتد عند القمة، وعطري الرائحة.

الزعفران غير معروف في البرية؛ ولكنه مستنبتٌ من اليونان. استُقدم الزعفران وزُرع على نطاق واسع في جميع أنحاء أوروبا والمغرب وتركيا وإيران وباكستان وغرب جبال الهيمالايا. قيل أن الزعفران زُرع في العراق، وأشار ماثيو في كتاب «نباتات العراق» (المجلد 8، ص 21) إلى أن المكان الذي يُدعى الزعفرانية ربما كان اسمه مُستوحى من أن النبات قد زُرع فيه من قبل.

القمم برتقالية اللون من محيط الغلاف الزهري والمياسم هي الجزء المُستخدم للأغراض الطبية وغيرها. وقد استُخدم الزعفران على نطاق واسع للأغراض الطبية منذ العصر اليوناني والروماني. في الحاضر، يُستخدم الزعفران بشكل رئيسي كعاملٍ مُنكّه يُستخدم في الطهي لا سيّما في المطبخ الإيراني، وكصبغة؛ بينما استُخدم منقوع الزعفران لتهدئة الحمى والصداع وعلاج الاكتئاب وكمهدئ وعلاج تضخم الكبد.

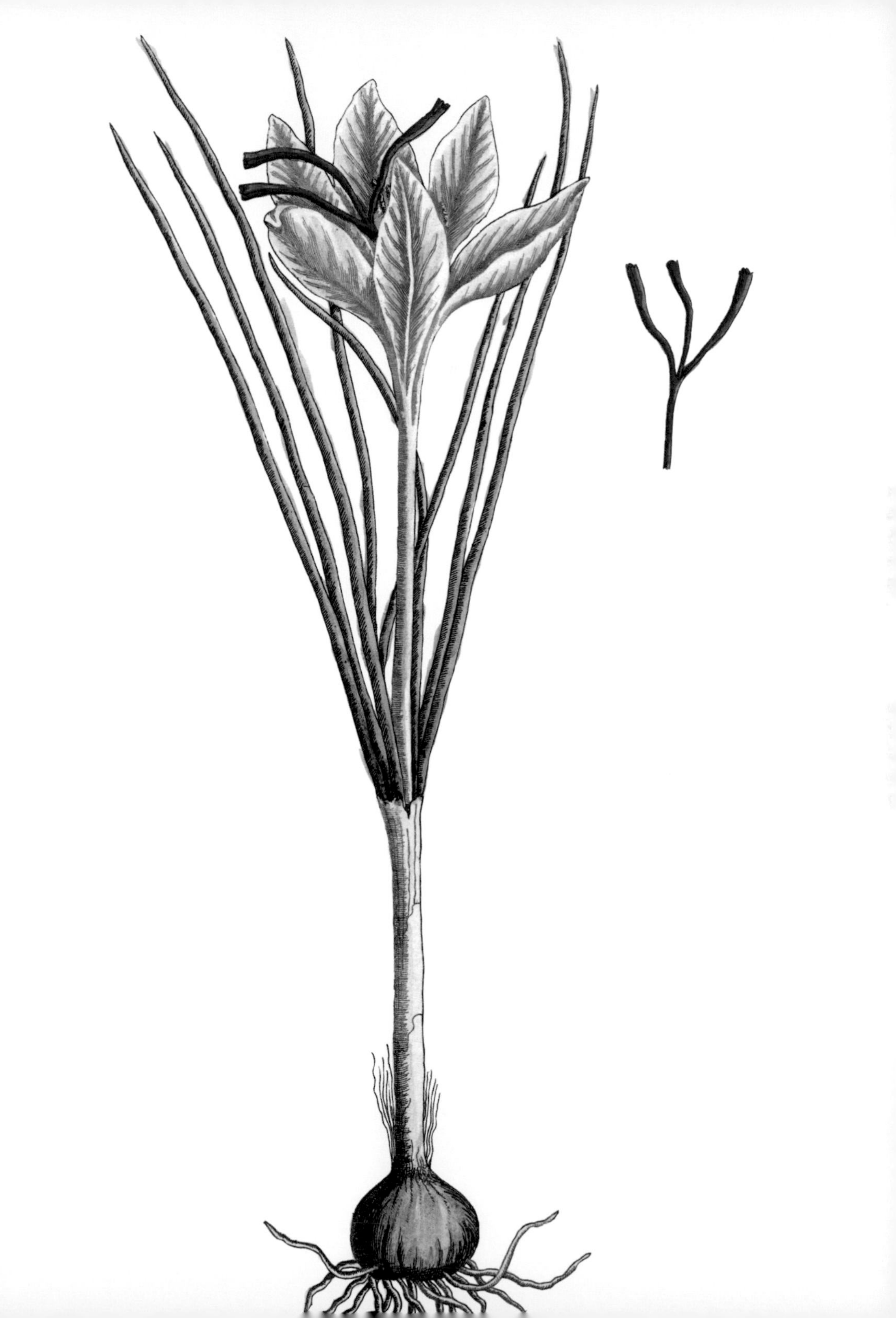

Cyperus rotundus L.

Family Cyperaceae

nut grass

الفصيلة السعدية

سعد *se'd* ;عشب الجوز

A creeping perennial 10 cm–1 m (4–39 in); culms 3-angled. Rhizomes long, covered with reddish-brown scales ending in a swollen, ellipsoid tuber (or 'nut'). Leaves glossy green. Inflorescence corymbose with the lateral flowers taller than central ones, simple to compound. Nutlet dark brown, obovoid, trigonous, minutely papillose.

Nut grass is native and widespread in tropical and subtropical regions of the Old World. In Iraq, it is common in the desert region, especially on irrigated alluvial plains and in the southern marsh district, where it is found on damp soil, sand and gravel, by irrigation channels, and as a weed among summer crops; alt. up to 900 m (3,000 ft); fl. & fr. May–Aug.

The use of the rhizomes of nut grass for medicinal purpose is recorded by early Greek and Muslim physicians. The small tubers are fragrant, resembling lemon and cardamom, and have been used for cleaning teeth and placed among clothes to keep away moths and other insects. Medicinally, the rhizomes have been used to regulate menstruation, treat uterine disorders and ulcers, and to prevent tooth decay. A decoction of the rhizomes has been used in Iraq as an anti-diarrheal, to decrease inflammation and as a sudorific. Tubers are used in East Africa for treating malarial fever. They have been prescribed for the treatment of stomach disorders and irritation of the bowels. Tubers are reported as a stomachic, emmenagogue and, when fresh, as a diaphoretic, diuretic and astringent. Powdered, dried or fresh tubers have been used as an insecticide in the Arabian Peninsula. An ointment made of the powdered tubers has been used on bee stings and bites and rubbed on breasts as a galactagogue. In Yemen the powdered rhizomes have been used to perfume clothes and hair.

نبات مُعمر زاحف. يبلغ طوله 01 سم - 1 م. وساقه مثلثة المقطع. الجذامير عمودية طويلة ثلاثية الزوايا ومغطاة بقشور بنية مائلة إلى الحمرة تنتهي بدرنات منتفخة بيضاوية الشكل (أو «تشبه الجوز»). الأوراق خضراء اللون لامعة. النورات عذقية الأزهار الجانبية فيها أطول من الأزهار المركزية، بسيطة إلى مركبة. الجوزات الصغيرة لونها بني داكن، بيضاوية مقلوبة، مثلثة المقطع، حليمية دقيقة.

الموطن الأصلي لعشب السعد وموطن انتشاره هما المناطق الاستوائية وشبه الاستوائية في العالم القديم. في العراق، يشيع وجود السعد في المنطقة الصحراوية، لا سيّما في السهول الرسوبية المروية وفي منطقة الأهوار الجنوبية، وينمو في التربة الرطبة والرملية والحصوية إلى جوار قنوات الري، كما ينمو كحشيشة بين المحاصيل الصيفية. ينمو النبات على ارتفاع يبلغ 900 م؛ موسم الإزهار والإثمار: مايو–أغسطس.

استخدم الأطباء اليونانيون والمسلمون الأوائل جذامير نبات السعد للأغراض الطبية. الدرنات الصغيرة عطرية تشبه الليمون والهيل (الحبهان). استُخدمت هذه الدرنات لتنظيف الأسنان وتوضع بين الملابس للحفاظ عليها من العث وغير ذلك من الحشرات. وقد استُخدمت الجذور لتنظيم الدورة الشهرية وعلاج اضطرابات الرحم والقروح، ولمنع تسوس الأسنان. استُخدم مستخلص الجذامير بالإغلاء في العراق كمضادٍ للإسهال ولتقليل الالتهابات وكمُعرِّق. تُستخدم الدرنات في شرق أفريقيا لعلاج حمى الملاريا، كما توصف لعلاج اضطرابات المعدة وتهيج الأمعاء. ورد أن الدرنات قد استُخدمت كمساعدٍ على الهضم، ومُدرّة للطمث. تُستخدم الدرنات الطازجة كمُعرِّقٍ ومُدرّة للبول وقابضة للمسام؛ بينما استُخدمت الدرنات المسحوقة أو المجففة أو الطازجة كمبيد حشري في شبه الجزيرة العربية. يُحضّر من الدرنات المسحوقة مرهمٌ يُدهن على لسعات النحل ولدغاته ويُستخدم في تدليك الثديين لإدرار اللبن. في اليمن، استُخدم مسحوق الجذامير لتعطير الملابس والشعر.

Eminium spiculatum (Blume) Schott

Family Araceae

الفصيلة اللوفية

laiyā ليا

A low-growing perennial herb with a depressed globose tuber giving rise to a thick peduncle. Leaf petioles usually have purple spots. Leaves with the lateral lobes divided to narrowed segments, spirally twisted. Inflorescence appears in the centre of the rosette, strongly smelling of carrion. Spathe tube 6–8 cm (2–3 in), pale purplish in the lower part, purple-spotted on the inner surfaces; spathe oblong-ovate, 10–15 cm (4–6 in), outer surface often purple-spotted, the inner surface dark blackish-purple, wrinkled and leathery. Spadix yellowish or brown, with female flowers near the base separated from the male flowers by a space in which sterile male flowers are distributed, apical appendage of spadix 6–8 cm (2–3 in) long. Fruit a berry, 1-seeded.

Eminium spiculatum is native to Egypt and Sinai, Israel and Palestine west to Turkey and east to Iran. In Iraq, it is common in the lower hills and grassland, less so in the lower forest zone, found on mountain slopes, in fields and grassy places on the upper plains; alt. 350–1,000 m (1,150–3,300 ft); fl. & fr. Apr.–May.

Guest (1933) notes that this unpleasant-smelling plant is reputed to be poisonous when consumed (like many species in the family); however, it has been reported to have been used as an antiseptic to treat itch and mites in sheep and other livestock, and in making cheese, where a small part of the corm is boiled and added to curdled milk. Given its apparent toxicity, it should only be used with caution.

عشب مُعمر قليل الارتفاع ذو درنة شبه كروية ينمو منها سويق سميك. عادة ما تكون أعناق الأوراق ذات بقع أرجوانية. الأوراق والفصوص الجانبية مُقسمة إلى عقلات ضيقة، وملتوية حلزونيا. النورات تظهر في وسط الوردة، وتفوح منها رائحة نتنة قوية. غلاف الطلع أنبوبي يبلغ طوله 6–8 سم، الجزء السفلي منه أرجواني شاحب اللون، أما الأسطح الداخلية فلونها أرجواني مرقط. غلاف الطلع بيضاوي مستطيل الشكل، يبلغ طوله 10–15 سم، السطح الخارجي له مرقط أرجواني اللون، أما السطح الداخلي فلونه أرجواني داكن مائل للون الأسود. ملمسه متجعد يشبه الجلد. الطلع أصفر أو بني اللون. تنبت الأزهار الأنثوية بالقرب من القاعدة على نحو منفصل عن الأزهار الذكرية بمساحة تتوزع فيها الأزهار الذكرية العقيمة. للطلع زائدة قمية يبلغ طولها 6–8 سم. الثمرة تشبه التوت وتحتوي على بذرة واحدة.

الموطن الأصلي لنبات *Eminium spiculatum* (الليا) هو مصر وسيناء وإسرائيل وفلسطين غربًا إلى تركيا وشرقًا إلى إيران. في العراق، ينتشر النبات على التلال السفلية والأراضي العشبية، ويقل في منطقة الغابات المنخفضة، كما يوجد على المنحدرات الجبلية، وفي الحقول والبقاع العشبية في السهول العليا. ينمو النبات على ارتفاع 350–1000 م؛ موسم الإزهار والإثمار: إبريل–مايو.

أشار جيست (1933) إلى أن هذا النبات ذا الرائحة الكريهة قد أُشتهر بكونه نباتًا سامًا عند تناوله (شأنه في ذلك شأن العديد من الأنواع التي تنتمي للفصيلة ذاتها)، إلا أنه عُرف باستخدامه كمُطهر لعلاج الحكة والعث لدى الأغنام وغيرها من الماشية. وقد استُخدم في صناعة الجبن، حيث يُغلى جزءٌ صغيرٌ من القرم، ويضاف إلى اللبن الرائب لصنع الجبن. ونظرًا لسميته المؤكدة، ينبغي توخي الحذر عند استخدامه.

Glaucium corniculatum (L.) Curtis

Family Papaveraceae

الفصيلة الخشخاشية

red-horned poppy

ورد نيسان *ward nisan* ; خشخاش البحر

An annual or biennial with branched stems 30–90 cm (12–35 in) tall. Leaves form a basal rosette, deeply divided with oblong to triangular, lobed or toothed segments. Flowers yellowish or red, petals orbicular, with a dark red-brown blotch at base. Fruit a conspicuously long, slender capsule, opening from the top by 2 valves; seeds embedded in a spongy false septum.

Red-horned poppy is native to Macaronesia and the Mediterranean Basin east to Iran. It is common in the lower hills and northwest sector of the desert region of Iraq, found in the desert plains and hills in gravelly soils, clay and silt; alt. 100–700 m (330–2,300 ft); fl. Mar.–Apr.

The plant has been used as a tonic and soothing agent, and the roots as a sedative, as a purgative, and as a soothing agent. It is recorded to have been used to adulterate opium.

نبات حولي أو حؤولي ذو سيقان متفرعة يبلغ طولها 90–30 سم. تُشكّل الأوراق وردة قاعدية، مُقسَّمة تقسيمًا عميقًا إلى شرائح مستطيلة إلى مثلثة ذات فصوص أو أسنان. الأزهار صفراء أو حمراء اللون، ذات بتلات دائرية الشكل، وبقعة حمراء/ بنية داكنة عند القاعدة. الثمرة طويلة وواضحة تحتوي على علبية بذور نحيلة، تفتح من أعلى بصمامين. البذور داخل غشاء إسفنجي فاصل يحد الفتحات الميكروبية في البويضات.

الموطن الأصلي لنبات خشاش البحر (ورد نيسان) هو ماكارونيسيا وحوض البحر الأبيض المتوسط شرقًا إلى إيران. ينتشر النبات في التلال المنخفضة والقطاع الشمالي الغربي من المنطقة الصحراوية في العراق، كما يوجد في السهول والتلال الصحراوية في التربة الحصوية والطين والطمي. ينمو النبات على ارتفاع 700–100 م؛ موسم الإزهار: مارس - إبريل.

استُخدم النبات كعاملٍ مُنشطٍ ومُهدئٍ؛ بينما استُخدمت الجذور كعاملٍ مُسكّنٍ ومُلينٍ ومُهدئٍ. تشير الأبحاث إلى استخدام النبات في غش الأفيون.

Glycyrrhiza glabra L.

Family Fabaceae

الفصيلة البقولية

liquorice

عرق السوس *irq as-sūs*

A perennial herb with underground stems (rhizomes), sticky to touch. Leaves divided into 9–17 leaflets, arranged in pairs along a central axis, with a single leaflet at the end. Flowers light blue to pale purple (rarely white), pea-like in shape. Pods reddish-brown, 2–5-seeded; seeds brown to blackish.

Liquorice is native to Eurasia, northern Africa and western Asia; it has been introduced into many countries and cultivated as a crop, particularly in Russia, Spain and the Middle East. In Iraq, it is common in the lower forest zone, on mountain slopes and valleys, in the lower hills along field margins, edges of rivers and streams, and on alluvial plains, in shady orchards and date gardens; alt. up to 1,200 m (3,900 ft); fl. & fr. June–Sept. on mountains; Apr.–May on plains.

Liquorice is best known for its use in confectionery, in the production of cough mixtures and throat lozenges and as an additive to medicines as a sweetener. It was known in ancient Mesopotamia, ŠŪŠU(M) (Akkadian) for uses of its roots and rhizomes. It has been used in medicine for more than 4,000 years, to treat sore throat, mouth ulcers, stomach ulcers, inflammatory stomach conditions and indigestion. The earliest record of the use of liquorice in medicine is in the code Humnubari (2100 BCE). It was also one of the important plants mentioned in Assyrian herbal documents (2000 BCE). Hippocrates (400 BCE) mentioned its use as a remedy for ulcers and quenching of thirst. Liquorice as a drug was also mentioned by Theophrastus and Dioscorides.

In Iraq, the root was once sold widely and exported. According to the *Flora of Iraq*, there was a liquorice factory and packing station in Makina, south Basra. Ground rhizome added to a cup of warm water has been used for the treatment of intestinal inflammation, kidney problems and abdominal pain. Liquorice rhizome chewed or made into tea with other anti-spasmodic herbs has often been taken for menstrual cramps.

عشب مُعمر تنمو سوقه تحت سطح الأرض (جذامير)، لزج الملمس. الأوراق مُقسَّمة إلى 9–17 وريقة صغيرة ومنتظمة في شكل أزواج بطول محور مركزي، وتوجد وريقة صغيرة واحدة في النهاية. يتدرج لون الأزهار من الأزرق الفاتح إلى الأرجواني الباهت (نادرًا ما تكون بيضاء) تشبه البازلاء في شكلها. الجريبات بنية اللون مائلة إلى الحمرة، تحتوي على 2–5 بذور يتروح لونها من البني إلى الأسود.

الموطن الأصلي لنبات عرق السوس هو أوراسيا وشمال أفريقيا وغرب آسيا. وقد استُقدم النبات إلى العديد من البلدان وزُرع كمحصول، لا سيّما في روسيا وأسبانيا والشرق الأوسط. في العراق، ينتشر النبات في منطقة الغابات المنخفضة، على المنحدرات والوديان الجبلية، وفي التلال السفلية بامتداد أطراف الحقول، وعلى حواف الأنهار والجداول، وفي السهول الرسوبية، وفي البساتين الظليلة ومزارع النخيل. ينمو النبات على ارتفاع يبلغ 1200 م؛ موسم الإزهار والإثمار على الجبال: يونيو – سبتمبر، وفي السهول: إبريل – مايو.

يشتهر عرق السوس باستخدامه في صناعة الحلويات، وفي إنتاج أخلطة لعلاج السعال وأقراص الحلق ويُضاف للأدوية كمُحلي. عُرف عرق السوس ŠŪŠU(M) (باللغة الأكادية) في بلاد ما بين النهرين قديمًا لاستخدام جذوره وجذاميره. وقد استُخدم النبات لأغراض طبية لأكثر من 4000 عام لعلاج التهاب الحلق وتقرحات الفم وقرحات المعدة وحالات التهاب المعدة وعسر الهضم. يرجع أقدم سجِّلٍ لاستخدام عرق السوس لأغراض طبية إلى قوانين حامورابي (2100 ق.م.)، كما كان من النباتات المُهمة المذكورة في الوثائق العشبية الآشورية (2000 ق.م.). أشار أبقراط (400 ق.م.) إلى استخدامه كعلاجٍ للقرحات وإخماد العطش، كما أشار كلٌ من ثاوفرسطس وديسقوريدوس إلى استخدام عرق السوس كدواء.

في العراق، كان الجذر يُباع ويُصدَّر على نطاق واسع. ووفقًا لكتاب «نباتات العراق»، كان هناك مصنعٌ لإنتاج عرق السوس ومحطة تعبئة في جنوب البصرة. يُضاف الجذمور المطحون إلى كوب من الماء الدافئ لعلاج التهاب الأمعاء واعتلالات الكلى وآلام البطن. غالبًا ما تُمضغ جذور عرق السوس أو توضع في الشاي، مع غيرها من الأعشاب المضادة للتشنج، لعلاج تقلصات الدورة الشهرية.

Inula helenium L.

Family Asteraceae

الفصيلة النجمية

elecampane

aqir عقير; *qir'he* قرعة

A robust perennial herb with a thick sub-tuberous rhizome. Stems several, erect 1–2 m (3.3–6.6 ft) tall. Basal leaves lanceolate-elliptic, with serrate margins. Inflorescence with few to many capitula. Involucre (base of flowerhead) hemispherical, frequently subtended by small bract-like leaves. Outer phyllaries ± densely velvety-tomentose; innermost subglabrous and membranous. Achene with pappus.

Elecampane is native to temperate Eurasia. It is widespread in Europe and widely naturalised from cultivation as a medicinal plant. It is found scattered in the mountain regions of Iraq, on the lower slopes of high mountains and valleys; alt. 1,000–1,200+ m (3,300–3,900 ft); fl. & fr. Aug.–Sep.

The roots have been used medicinally since antiquity to cure a shortness of breath, and as a sedative, antispasmodic and tonic. It was mentioned by Ibn al-Baitar (c. 1240 CE) for strengthening the bladder and as a diuretic. In Iraq, the powdered roots mixed with honey have been used as an expectorant and for the treatment of infertility (in both men and women).

عشب مُعمر قوي ذو جذمور سميك شبه درني. للعشب عدة سوق منتصبة يبلغ طول كل منها 2–1 م. الأوراق قاعدية رمحية بيضاوية الشكل، ذات حواف مسننة. النورات لها عدد قليل إلى كثير من الرؤيسات. القناب (قاعدة الزهرة-الرأس) نصف كروي، تقابله في الغالب أوراق صغيرة تشبه القنابات. الأوراق النباتية الخارجية ± مخملية-وبرية كثيفة. السطح الداخلي شبه أملس رقيق. الثمرة ذات زغب.

الموطن الأصلي لنبات *Inula helenium* (العقير) هو أوراسيا المعتدلة، وينتشر العشب في أوروبا ويُزرع بوصفه نبات طبي مستوطن على نطاق واسع. يوجد العشب متفرقًا في المناطق الجبلية في العراق، ويوجد على المنحدرات المنخفضة للجبال العالية والوديان. ينمو النبات على ارتفاع +1200–1000 م؛ موسم الإزهار والإثمار: أغسطس – سبتمبر.

استُخدمت الجذور طبيًا منذ القدم لعلاج ضيق التنفس، وكمُسكّنٍ ومُضاد للتشنج ومُنشطٍ. وقد أشار إليه ابن البيطار (نحو عام 1250 م) لتقوية المثانة وإدرار البول. في العراق، استُخدم مسحوق الجذور الممزوج بالعسل كطاردٍ للبلغم ولعلاج العقم (عند الرجال والنساء).

Lysimachia arvensis (L.) U. Manns & Anderb.

Family Primulaceae

الفصيلة الربيعية

Anagallis arvensis L.
scarlet pimpernel

الزريقاء; *'ain al jamal* عين الجمل

An annual herb, often with much-branched ascending stems, up to 50 cm (20 in) long, 4-winged. Leaves bright green, opposite, elliptic to ovate, sessile. Flowers solitary, axillary, orange-red or blue, on long filiform pedicels initially 1–4 cm (0.4–1.6 in) and erect, later elongating and recurving in fruit. Filaments of stamens bearded. Capsule many-seeded.

Scarlet pimpernel is native to Europe, west to western Asia to Socotra, and across North Africa south to Ethiopia; it has been introduced and is widespread in the temperate regions of both hemispheres. In Iraq, it is found throughout in the lower mountains and plains, and is common in cultivated fields, irrigated places and gardens, on silty, loamy and sandy/gravelly soils; alt.: 50–180 m (165–590 ft); fl. Mar.–Apr.

In Iraq, a decoction of the whole plant has been used to treat mental problems, inflammation of the brain and ulcers. In the Arabian Peninsula, it has been used to treat skin rash and ulcers, snake bites and epilepsy.

عشبة حولية ذات سيقان صاعدة كثيرة التفرعات، يبلغ طولها 05 سم، وذات 4 أجنحة. الأوراق خضراء زاهية اللون، متعاكسة، إهليلجية بيضاوية الشكل، لا عنقية. الأزهار منفردة، إبطية، برتقالية - حمراء أو زرقاء. توجد الأزهار على سويق خيطي منتصب طويل يبلغ طوله مبدئيًا 1-4 سم، ثم يمتد السويق وينحني حاملًا الثمرة. خيوط السداة مشعرة. عليبة البذور تحتوي على كثير من البذور.

الموطن الأصلي لنبات عين الجمل (أو حشيشة العلق/ اللبين الحلقي) هو أوروبا، غرب آسيا إلى سقطرى في اليمن، وعبر شمال أفريقيا جنوبًا إلى إثيوبيا؛ وقد استُقدمت العشبة وانتشرت في المناطق المعتدلة في نصفي الكرة الأرضية. في العراق، توجد العشبة في جميع أنحاء الجبال والسهول المنخفضة، كما تشيع في الحقول المزروعة والأماكن المروية والحدائق، في التربة الطينية والطفلية والرملية / الحصوية. ينمو النبات على ارتفاع: 50–180 م؛ موسم الإزهار: مارس - إبريل.

في العراق، استُخدم مستخلص النبات بأكمله بالإغلاء لعلاج الاعتلالات العقلية والتهاب الدماغ وعلاج التقرحات. في شبه الجزيرة العربية، استُخدم النبات بالكامل لعلاج الطفح الجلدي والتقرحات ولدغات الثعابين وعلاج الصرع.

Marrubium vulgare L.

Family Lamiaceae

الفصيلة الشفوية

white horehound,
common horehound

alfar asiyon al abiyad الفراسيون الأبيض;
alfar asiyon alshai' الفراسيون الشائع;
heshshat al-kalb حشّشت الكب

A perennial herb up to 50 cm tall (20 in). Leaves grey-green, broadly ovate, with crenate margins, white-hairy. Inflorescence with dense whorls of white flowers at nodes, subtended by leafy bracts. Calyx tubular with 10 spreading hooked teeth; corolla 2-lipped.

Native to Macaronesia, Europe and from the Mediterranean Basin to the western Himalaya; introduced in the New World and Australasia.

Marrubium vulgare has been cultivated and grown as a medicinal herb since antiquity. In Iraq, an infusion made from the leaves and flowering tops has been recorded to treat the common cold, and as an expectorant, diuretic, carminative and restorative.

عشب مُعمر يبلغ طوله ٥٠ سم. الأوراق رمادية مخضرة، بيضاوية الشكل عريضة، وحوافها ذات أسنان مستديرة، بيضاء مزغبة. النورات ذات جدلات كثيفة تحتوي على أزهار بيضاء عند العقد يقابلها قنابات مورقة. الكأس أنبوبي ذو ١٠ أسنان معقوفة منبسطة، وتويج ذو شفتين.

الموطن الأصلي للعشب هو ماكارونيسيا وأوروبا ومن حوض البحر الأبيض المتوسط إلى غرب جبال الهيمالايا، وقد استُقدم العشب إلى العالم الجديد والأسترالاسيا.

لطالما زُرع نبات *Marrubium vulgare* (حشيشة الكلب) كعشبٍ طبيٍّ منذ القدم. في العراق، سُجِّل منقوع الأوراق والقمم المزهرة لعلاج نزلات البرد، وكطاردٍ للبلغم ومُدرٍّ للبول وطاردٍ للريح ومُقوٍّ.

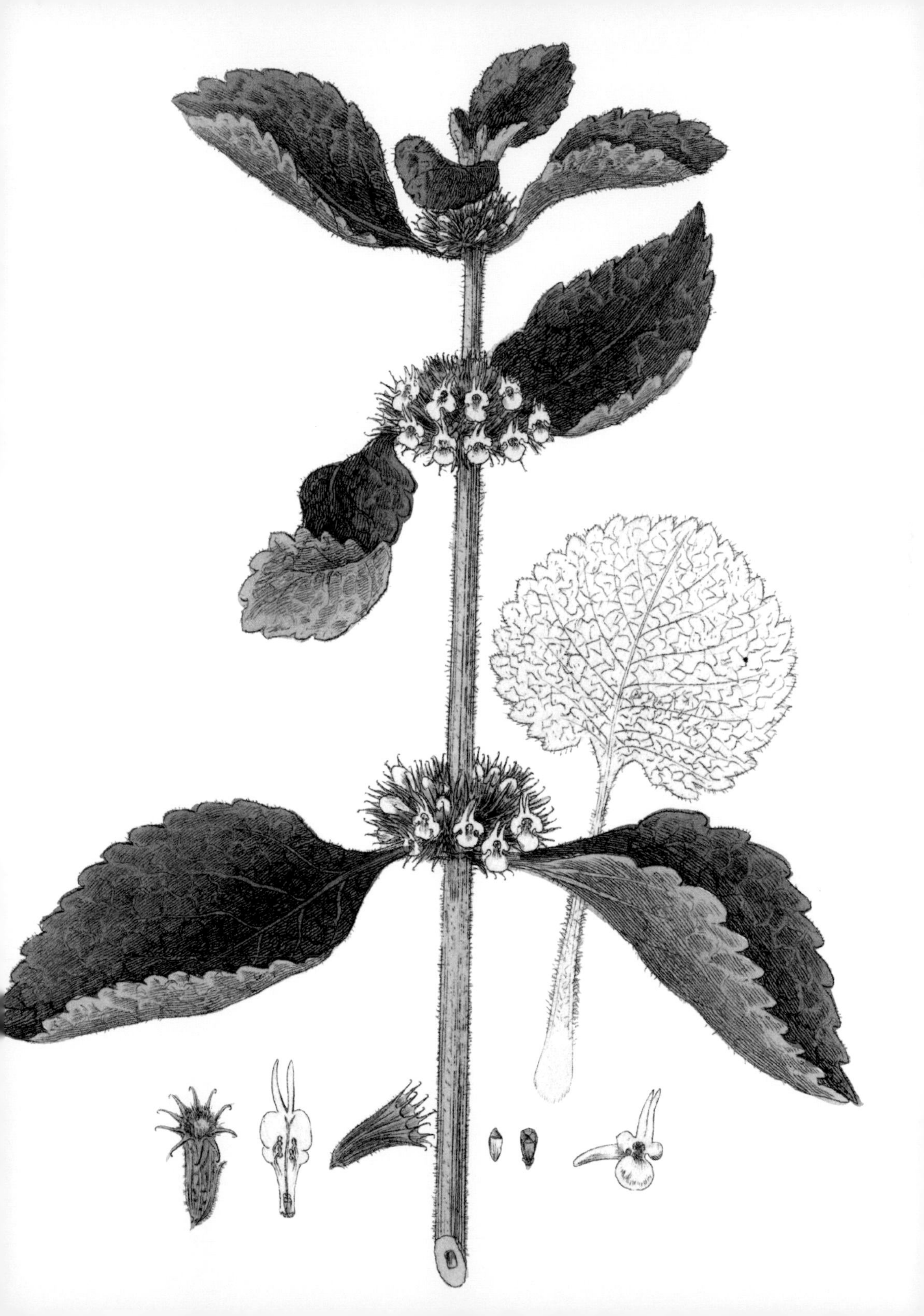

Matricaria chamomilla L.

Family Asteraceae

الفصيلة النجمية

chamomile

بابونج *babunnej* ;بيبون *beybūn* ;أقحوان

A glabrous annual herb, 10–60 cm (4–24 in) tall, branched from the base. Leaves divided into linear segments. Flowers borne in capitula, 1.5–2.0 cm (0.6–0.8 in) in diameter, with ray florets white and the disc florets golden yellow. Achenes taper towards the base with pappus forming a corona.

Chamomile is native to temperate Europe and Asia to China; it is often cultivated. In Iraq, it is found scattered on the Mesopotamian alluvial plains, in waste places and near cultivation; alt. 50–250 m (160–820 ft); fl. Apr.–June.

Chamomile has a long history of use in medicine. An infusion of the flowers has been used for treating coughs, shortness of breath, the common cold, asthma, flatulence and abdominal pain. It has also been applied on facial discolourings, skin burns and to improve damaged hair. The infusion has also been used to treat jaundice.

عشبة حولية مجعدة متفرعة من قاعدتها يبلغ طولها 10–60 سم. الأوراق مُقسَّمة إلى فصوص طولية. الأزهار محمولة في رؤيسات، يبلغ قطر كل منها 1.5–2.0 سم، مع زهيرات بيضاء كثيرة التفرعات وزهيرات قرصية صفراء ذهبية. الثمرة مدببة ذات زغب تشبه التاج في شكلها.

الموطن الأصلي للأقحوان (البابونج) هو أوروبا وآسيا إلى الصين؛ حيث يُزرع في الغالب. في العراق، يوجد النبات على نحو متفرق في السهول الرسوبية لبلاد ما بين النهرين، في الأراضي المهملة وبالقرب من المزارع. ينمو النبات على ارتفاع 50–250 م؛ موسم الإزهار: إبريل - يونيو.

لنبات البابونج تاريخ طويل من الاستخدام لأغراض طبية؛ حيث استُخدم منقوع الأزهار للسعال وضيق التنفس وعلاج نزلات البرد والربو وانتفاخ البطن وآلام البطن، كما استُخدم لعلاج التصبغات السوداء على الوجه وحروق الجلد ولتحسين الشعر التالف. وقد استُخدمت منقوعات الأزهار لعلاج اليرقان.

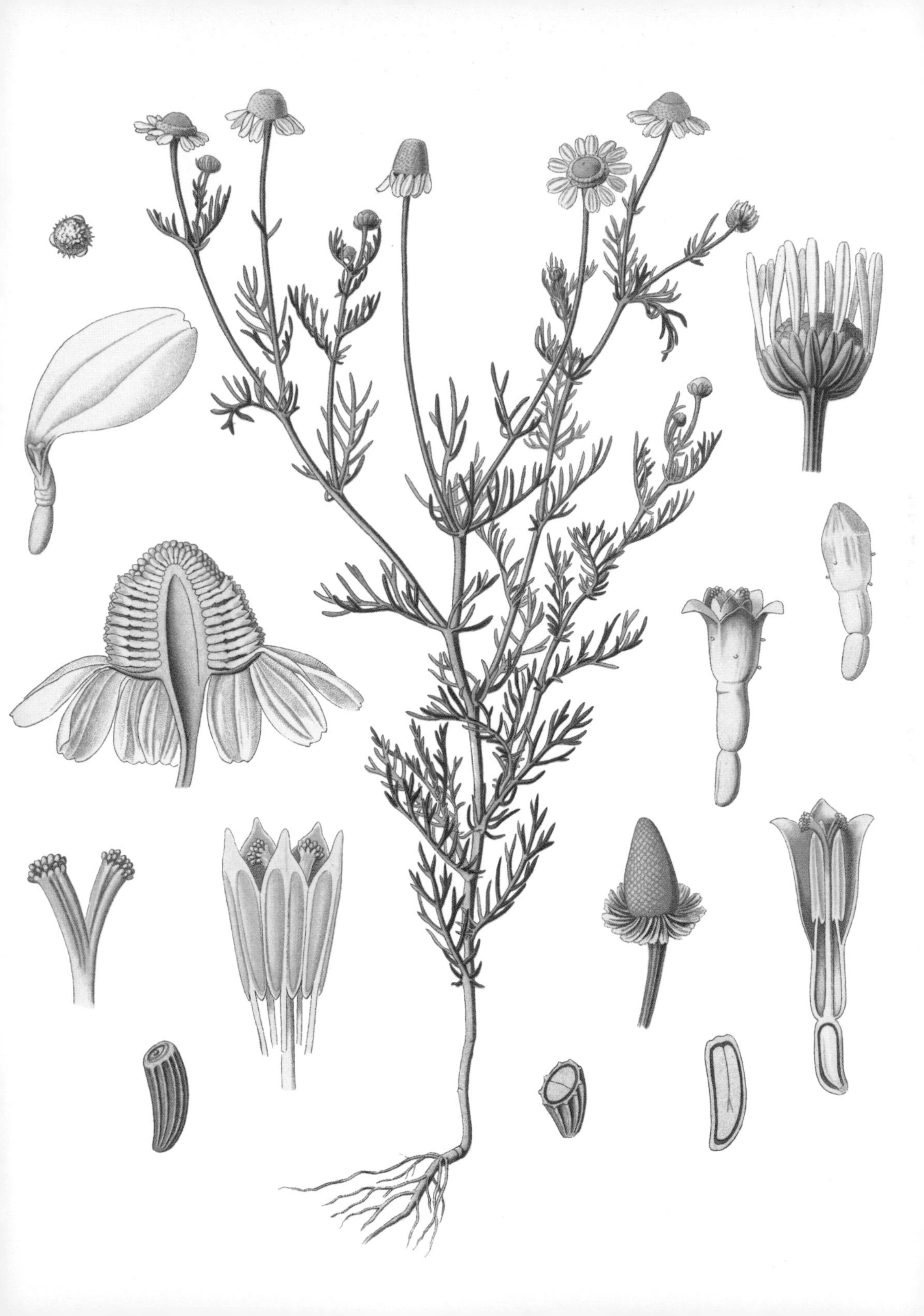

Myrtus communis L.

Family Myrtaceae

الفصيلة الآسية

myrtle

ياس *yās*, اس *ās*; ميرتل

An evergreen shrub or a small tree, to 2 m (6.6 ft). Leaves opposite, bright green, ovate to elliptic, aromatic when crushed. Flowers white, about 20 mm (0.8 in) across, solitary to few, borne in axillary cymes. Petals and sepals reflexed when flower fully open. Stamens many, well exserted. Berries ovoid, 7–10 mm (0.3–0.4 in), red, ripening blue-black.

Native from Macaronesia and the Mediterranean Basin east to Pakistan; introduced in India and southern Africa. Cultivated widely in Iraq for its fragrance, edible fruits and uses in traditional medicine. It is found naturalised in the lower mountain valleys, among oak forest, often by water; alt. 600–1,300 m (2,000–4,300 ft); fl. & fr. May–June.

Myrtle has been used as an aromatic and as an ingredient in perfume for ritual offerings in ancient cultures of the Middle East, India and southern Europe. Known as ASU(M) or ASSU (Akkadian) in ancient Mesopotamia, the earliest documentation of myrtle as a ritual plant is from Sumer, found in cuneiform text from the legendary tale of Gilgamesh used in sacrifice to please the Gods: '*I poured reeds, cedar, and myrtle ... the gods smelled the savour*' (c. 2100 BCE, Tablet XI). Myrtle has also been used in medicinal and magical preparations since antiquity. It was cited by Ibn al-Baitar (c. 1240 CE) as a treatment for urinary problems. In Ancient Eastern and Mediterranean regions, myrtle was an important plant in wedding ceremonies, and was considered to be a potent symbol of love and fertility. It was used by Ancient Greeks to respect the dead; graves were decorated with myrtle in the Mediterranean.

In Arabia, myrtle has been used medicinally to treat abdominal colic, coughs, fevers and headache, and powdered leaves have been applied to blisters, stings and ulcers. Leafy twigs placed among clothes serve as perfume, give a pleasant smell to clothes and are said to protect against moths.

شجيرة دائمة الخضرة أو شجرة صغيرة يبلغ طولها 2 م. الأوراق متعاكسة، خضراء زاهية اللون، بيضاوية إهليلجية، تصدر عنها رائحة عطرية عند سحقها. الأزهار بيضاء اللون، يبلغ عرضها نحو 20 مم، ما بين منفردة إلى قليلة. الأزهار محمولة في نورات محدودة إبطية. تنحني البتلات والسبلات عندما تتفتح الزهرة تمامًا. الأسدية كثيرة وناتئة نتوءًا واضحًا. الثمرة توتية الشكل بيضاوية حمراء اللون، يبلغ طولها 10–7 مم. عند نضج الثمرة، يتحول لونها إلى أزرق-أسود.

الموطن الأصلي للنبات هو ماكارونيسيا وحوض البحر الأبيض المتوسط شرقًا إلى باكستان، وقد استُقدم إلى الهند وجنوب أفريقيا. يُزرع على نطاق واسع في العراق لرائحته وثماره الصالحة للأكل واستخداماته في الطب التقليدي. يستوطن النبات الوديان الجبلية المنخفضة، وبين غابات البلوط، وفي كثير من الأحيان على حافة المسطحات المائية. ينمو النبات على ارتفاع 1300–600 م؛ موسم الإزهار الإثمار: مايو - يونيو.

استُخدم نبات الميرتل (الآس) كمادةٍ عطريةٍ وكمُكوِّنٍ في العطور المُستخدمة في العروض الطقوسية الخاصة بالثقافات القديمة في الشرق الأوسط والهند وجنوب أوروبا. عُرف النبات باسم (M) ASSU أو ASSU (في اللغة الأكادية) في بلاد ما بين النهرين القديمة. يرجع أقدم توثيق لنبات الآس كنباتٍ طقوسيٍّ إلى الحضارة السومرية؛ حيث وجد نصٌّ مسماريٌّ يحكي الحكاية الأسطورية للملك جلجامش الذي كان يستخدمه كقربانٍ لإرضاء الآلهة «لقد سكبت القصب والأرز والآس ... وقد شمَّ الآلهة الرائحة» (نحو عام 2100 ق.م.، اللوح الحادي عشر). استُخدم الآس في المستحضرات الطبية ولأغراض السحر منذ القدم. وقد استشهد به ابن البيطار كعلاجٍ لاعتلالات المسالك البولية. في مناطق الشرق والبحر الأبيض المتوسط قديمًا، كان نبات الآس نباتًا مُهمًا في حفلات الزفاف، وكان يُعتبر رمزًا قويًا للحب والخصوبة. وقد استخدم اليونانيون القدماء النبات احترامًا للموتى؛ حيث اعتادوا تزيين القبور به في البحر الأبيض المتوسط.

في شبه الجزيرة العربية، استُخدم الآس في الأغراض الطبية لعلاج المغص البطني ونوبات السعال والحمى والصداع. وقد استُخدم مسحوق الأوراق على البثور واللدغات والقروح. توضع الأغصان الورقية بين الملابس لتعطيرها وإعطاء رائحة طيبة لها، كما يقال إنها تحمي من العث.

Nigella sativa L.

Family Ranunculaceae

black cumin, black seed

الفصيلة الحوذانية

حبة سوداء *habbat sōdā*; الكُمون الأسود

An annual, erect herb to 50 cm (20 in). Leaves dissected into narrow segments. Flowers solitary, terminal. Sepals conspicuous, blue; petals 5–8, greenish-yellow, smaller than the sepals, 2-lobed, swollen at base, nectariferous. Fruit formed of follicles, fused at the apex to form a capsule, each follicle terminating in a long beak. Seeds trigonous, black, finely rugulose.

Black cumin is native from Eastern Europe (Bulgaria and Romania) east to Turkey, Iraq and Iran. In Iraq, it is rarely found in the wild although extensively cultivated and a frequent garden escape. In the wild, it is found on rocky mountains, gravel desert plains and river banks; alt. 250–600 m (800–2,000 ft); fl. & fr. Jan.

Seeds of black cumin have a long history of use in Europe and Asia for culinary and medicinal purposes. Seeds are often sprinkled on bread and other foods. In the Hadith, Abu Hurayrah narrates that Prophet Mohammed said of the seeds of *Nigella sativa*: إنَّ فِي الْحَبَّةِ السَّوْدَاءِ شِفَاءً مِنْ كُلِّ دَاءٍ إلاَّ السَّامَ 'in the black seeds there is remedy (cure) for every disease except death'. The Arabic name *habbat sōdā* is mentioned by Ibn al-Baitar (c. 1240 CE), referring to the seeds being used to treat fever, and as a carminative, digestive and tonic. In Iraq the seeds have been used as a stimulant, carminative and diuretic, and externally as an application on skin eruptions. Mati & de Boer (2011) document the use of seeds for diabetes and hypertension, for pneumonia, tonsillitis, blood circulation and cancer and to boost the immune system. In Arabia (Oman), the seeds are soaked in rose oil for use as eye-drops, they are eaten to relieve congestion and difficult breathing and flatulence. Crushed seeds mixed with ginger and oil have been used on polio-affected limbs to induce blood circulation. In Yemen the seeds have been used for constipation and haemorrhoids and worn in amulets to protect against evil spirits.

عشب حولي منتصب يبلغ طوله 50 سم. الأوراق مُقسَّمة إلى مقاطع ضيقة. الأزهار منفردة طرفية. السبلات واضحة زرقاء اللون. يوجد 8–5 بتلات لونها أصفر مائل إلى الخضرة. البتلات أصغر حجمًا من السبلات ولها فصين. البتلات منتفخة من عند القاعدة، ورحيقية. تتكوّن الفاكهة من بصيلات، ملتحمة من عند القمة لتكون علبية بذور؛ حيث تنتهي كل بصيلة بمنقار طويل. البذور مثلثة الشكل سوداء اللون متجعدة جدًا.

الموطن الأصلي للحبة السوداء هو أوروبا الشرقية (بلغاريا ورومانيا) شرقًا إلى تركيا والعراق وإيران. في العراق، نادرًا ما يوجد النبات في البريّة، ولكنه يُزرع على نطاق واسع ويكثُر في الحدائق. ينمو النبات في البريّة على الجبال الصخرية والسهول الصحراوية الحصوية وضفاف الأنهار. ينمو النبات على ارتفاع 250–600 م؛ موسم الإزهار والإثمار: يناير.

لطالما استُخدمت بذور الحبة السوداء في أوروبا وآسيا لأغراض الطهي والأغراض الطبية. غالبًا ما تُنثر البذور على الخبز والأطعمة الأخرى. وقد ورد في الحديث الشريف، روي أبو هريرة من حديث النبي (صلّ الله عليه وسلم) عن بذور حبّة البركة: «إنَّ فِي الْحَبَّةِ السَّوْدَاءِ شِفَاءً مِنْ كُلِّ دَاءٍ إلاَّ السَّامَ» وذكر ابن البيطار (نحو عام 1250 م) الاسم العربي «الحبّة السوداء» للإشارة إلى البذور المُستخدمة في علاج الحمى، وكطاردٍ للريح ومُهضمٍ ومُنشطٍ. في العراق، استُخدمت البذور كمُنبهٍ، وطاردٍ للريح، ومُدرٍّ للبول، واستُخدمت خارجيًا على الطفح الجلدي. وقد وثّق ماتي ودي بوير استخدام البذور لعلاج داء السكري وارتفاع ضغط الدم والالتهاب الرئوي والتهاب اللوزتين والدورة الدموية والسرطان ولتقوية جهاز المناعة. وفي شبه الجزيرة العربية (عمان)، تُنقع البذور في زيت الورد ويُستخدم الخليط كقطرة للعين، أو تؤكل لتخفيف الاحتقان وصعوبة التنفس وانتفاخ البطن. استُخدمت البذور المطحونة الممزوجة بالزنجبيل والزيت على الأطراف المصابة بشلل الأطفال لتحفيز الدورة الدموية. وفي اليمن، استُخدمت البذور لعلاج الإمساك والبواسير ووضعت في التمائم للحماية من الأرواح الشريرة.

The Prophet Mohammed (PBUH) made specific statements on 65 medicinal plants, of which the statement on *Nigella sativa* (black cumin, black seed) is the most important.

وقد ذكر النبي محمد (صلّ الله عليه وسلم) 65 نباتًا طبيًا على وجه التحديد، أهمها نبات *Nigella sativa* (المعروف باسم الكُمون الأسود أو حبّة البركة).

Oxalis corniculata L.

Family Oxalidaceae

الفصيلة الحُمّاضِيّة

Indian sorrel

حميض *hummaidh* ; حميض هندي

An annual herb with procumbent or ascending stems, usually rooting below. Leaves with 3 leaflets, the lateral leaflets smaller than the terminal one, all leaflets with a tapering base and a notched apex. Flowers yellow, with petals tapering at base. Capsule oblong-cylindrical, 5-angled.

Native to South and Central America from Mexico to Venezuela, Peru and the Caribbean. Indian sorrel has been introduced and naturalised almost throughout the tropics, subtropics and other warmer parts of the world. In Iraq, it is mainly found around Baghdad and in irrigated areas of the Mesopotamian plain in cultivated areas, date orchards, small gardens, on clay soil, and by irrigation canals; alt. 10–770 m (33–2,500 ft); fl. Nov.–May.

The vernacular name *hummaidh* refers to the sour taste of the leaves and stems. The leaves have been used as an astringent and diuretic, to increase appetite and as a detoxifying agent.

عشب حولي ذو سوق افتراشية أو صاعدة، وعادة ما تتجذر في الأسفل. الأوراق بها 3 وريقات صغيرة. الوريقات الجانبية أصغر حجمًا من الوريقات الطرفية. جميع الوريقات ذات قاعدة مستدقة وقمة مسننة. الأزهار صفراء اللون ذات بتلات مستدقة من عند القاعدة. عليبة البذور مستطيلة - أسطوانية، مخمسة الزوايا.

الموطن الأصلي للنبات هو أمريكا الجنوبية والوسطى من المكسيك إلى فنزويلا وبيرو ومنطقة البحر الكاريبي. استُقدم الحميض الهندي إلى جميع أنحاء المناطق الاستوائية وشبه الاستوائية والمناطق الأكثر دفئًا من العالم تقريبًا واستوطنها. في العراق، يوجد النبات بشكل رئيسي حول بغداد وفي المناطق المروية في سهل بلاد ما بين النهرين في المناطق المزروعة وبساتين النخيل والحدائق الصغيرة. ينمو النبات في التربة الطينية وعلى جوانب قنوات الريّ. ينمو النبات على ارتفاع 10–770 م؛ موسم الإزهار: نوفمب – مايو.

يشير الاسم الدارج، حميض، إلى الطعم الحامض للأوراق والسوق. وقد استُخدمت الأوراق كعاملٍ قابضٍ للمسام ومُدرٍّ للبول وفاتحٍ للشهية ومُزيلٍ للسموم.

Peganum harmala L.

Family Nitrariaceae

wild rue; Syrian rue

الفصيلة الحرملية

حرمل *harmal*; حرمل بريّ; حرمل شامي

Perennial herb to 60 cm (24 in) tall, glabrous throughout. Leaves divided into linear segments. Flowers white, solitary or in a loose corymb inflorescence, the petals streaked with green or yellow. Fruit depressed globose, 6–10 mm (0.2–0.4 in) in diameter, with three or four distinct bulges.

Wild rue is native from the Mediterranean Basin and North Africa east to Mongolia and India. It is common on the lower hills and desert region of Iraq, on sandy and gravel soils, in waste and disturbed spaces, near ancient ruins and graveyards, along wadis in the desert, and as a ruderal in disturbed places and around human settlements; alt. up to 300 m (980 ft); fl. & fr. Mar.–May. Sep.–Nov.

The plant and seeds were used medicinally by the Greeks and Romans, and were noted in European herbals of the seventeenth century. The seeds are exported from Iran to India, where the plant was originally introduced by Muslims as a medicinal herb.

In Iraq the plant has been used to treat ligament, joint and back pain. It is also used to dispel curses and spells and as incense to kill germs. The custom of sprinkling the seeds on burning coals at marriage ceremonies to avert the influence of the 'evil eye' has prevailed in Iraq. Burning the seeds and inhaling the fumes has been a ritual performed to dispel evil spells and germs. Smoke from burning seeds is said to drive away epidemics. The seeds were reputed to have a purifying property and to stimulate the sexual system. The use of seeds as a diuretic was cited by Ibn al-Baitar (c. 1240 CE). According to Mati & de Boer (2011), in Iraq half a teaspoon of ground seeds is eaten after food or an infusion of ground seeds is taken for a week to treat ligament, joint and back pain.

عشب مُعمر، يبلغ طوله 60 سم، كله أملس. الأوراق مُقسَّمة إلى فصوص طولية. الأزهار بيضاء اللون منفردة أو في نورات عذقية فضفاضة. البتلات مخططة باللون الأخضر أو الأصفر. الثمرة شبه كروية، يبلغ قطرها 6–10 مم، ولها 3 أو 4 انتفاخات مميزة.

الموطن الأصلي للحرمل البريّ هو المنطقة من حوض البحر الأبيض المتوسط وشمال أفريقيا شرقًا إلى منغوليا والهند. في العراق، يشيع وجود النبات في التلال والمنطقة الصحراوية، وينمو على التربة الرملية والحصوية، وفي الأماكن المهملة والتربة المضطربة، بالقرب من الآثار والمقابر القديمة، وبامتداد الوديان في الصحراء، كما ينمو الحرمل عشوائيًا في التربة المُضطربة وحول التجمعات البشرية. ينمو النبات على ارتفاع يبلغ 300 م؛ موسم الإزهار والإثمار: مارس - مايو. سبتمبر - نوفمبر.

استخدم الإغريق والرومان النبات والبذور لأغراضٍ طبيةٍ وعُدَّ النبات ضمن الأعشاب الأوروبية في القرن السابع عشر. تُصدَّر البذور من إيران إلى الهند، حيث استقدم المسلمون النبات في الأصل كعشبةٍ طبيةٍ.

في العراق، استُخدم النبات لعلاج آلام الأربطة والمفاصل والظهر، كما استُخدم لفك اللعنات والتعاويذ وكبخورٍ لقتل الجراثيم. سادت في العراق عادة نثر البذور على الجمر المحترق في مراسم الزواج لتجنب أثر «الحسد». كان حرق البذور واستنشاق أبخرتها طقسًا يؤدى لفك تعاويذ الشر وطرد الجراثيم. ويُقال إن الدخان المتصاعد جرّاء حرق البذور يزيل الأوبئة. اشتُهرت البذور بخاصية التنقية وتحفيز الجهاز التناسلي. وقد استشهد ابن البيطار بالبذور كمُدرٍّ للبول، بينما أشار ماتي ودي بوير (2011) إلى تناول نصف ملعقة صغيرة من البذور المطحونة بعد الأكل أو تناول منقوع البذور المطحونة لمدة أسبوع كعلاجٍ لآلام الأربطة والمفاصل والظهر في العراق.

Phyla nodiflora (L.) Greene

Family Verbenaceae

الفصيلة اللويزية

turkey tangle, creeping vervain

bulaih بليحه

A prostrate herb, with slender, often purplish creeping stems 20–45 cm (8–18 in) long. Leaves green to purplish, margins serrate at apex. Inflorescence of cylindrical spikes to 2.5 cm (1 in) long; flowers pink or white with a yellow centre. Fruit obovoid.

Native to the tropics and subtropics and warm temperate regions of Africa. In Iraq it is found by ponds and lakes, marshes and in ditches, in rice fields and as a weed in cultivated land, in mud and silt; alt. up to 150 m (500 ft); fl. & fr. Apr.–Dec.

The herb has been used a coolant, diuretic and a febrifuge; it has also been used to relieve knee joint pain and as a poultice for boils and ulcers. Extracts of the leaves are recorded to have been used as an antibacterial.

عشب رعوي، ذو سوق زاحفة نحيلة غالبًا ما تكون أرجوانية اللون، يبلغ طولها 20–45 سم. الأوراق خضراء إلى أرجوانية، ذات حواف مسننة عند القمة. النورات ذات سنابل أسطوانية يبلغ طولها 2.5 سم. الأزهار وردية أو بيضاء اللون وصفراء في الوسط. الثمرة بيضاوية مقلوبة.

الموطن الأصلي للنبات هو المناطق الاستوائية وشبه الاستوائية والمناطق الدافئة المعتدلة في أفريقيا. يوجد النبات في العراق إلى جوار البرك والبحيرات والمستنقعات وفي الخنادق وفي حقول الأرز، وينمو كحشيشة في الأراضي المزروعة في التربة الطينية والطميية. ينمو النبات على ارتفاع يبلغ 150 م؛ موسم الإزهار والإثمار: إبريل - ديسمبر.

استُخدم هذ العشب كمُبرِّدٍ ومُدرٍّ للبول ومُلطفٍ للحمى، كما كان يُستخدم لتخفيف آلام مفصل الركبة، وككمادةٍ للدمامل والتقرّحات. سُجِّل مُستخلص الأوراق كمُضادٍ للبكتيريا.

Plantago ovata Forssk.

Family Plantaginaceae

الفصيلة الحملية

P. ispaghula

ispagula; spogel seeds

ربله *riblah*؛ زباد *zibad* بذور سبوجل؛ إسباجيولا

An annual, short-stemmed (or stemless) herb, usually to 10 cm (4 in) tall, branched at base. The whole plant covered with long white hairs. Leaves linear to narrowly lanceolate, 3-veined. Peduncle as long as leaves or longer, bearing a globose to ovoid spike, densely flowered. Flowers minute, white. Capsule with 2 ovoid seeds.

Native from southeastern Spain and North Africa eastwards to India and Somalia, southwest and south-central USA to New Mexico. In Iraq, common in desert and wadi beds, on sandy gypsophilous and calcareous soils; in the desert plateau and lower Mesopotamian region found on sandy patches, limestone and conglomerate hillsides and on hard silty plains; alt. 10–150 m (33–490 ft); fl. & fr. Feb.–June.

A decoction of the seeds has been used medicinally for treating chronic constipation, chronic bacillary dysentery and as a poultice. The plant is recorded as spring fodder in the Iraq–Nejd border areas.

عشب حولي قصير الساق (أو بلا ساق)، يبلغ طوله عادة ١٠ سم، متفرع من القاعدة. النبات بأكمله مغطى بشعر أبيض طويل. الأوراق خطية إلى رمحية ضيقة، ثلاثية التعريق. النبات ذو سويقات لها نفس طول الأوراق أو أطول وتحمل سنبلة كروية إلى بيضاوية الشكل، كثيفة الأزهار. الأزهار دقيقة بيضاء اللون. عليبة البذور تحتوي على بذرتين بيضاويتين.

الموطن الأصلي للنبات هو جنوب شرق إسبانيا وشمال أفريقيا شرقًا إلى الهند والصومال، وجنوب غرب وجنوب وسط الولايات المتحدة الأمريكية إلى شمال المكسيك. في العراق، ينتشر النبات في الصحارى والوديان، وفي التربة الرملية الجصية والجيرية، وفي الهضبة الصحراوية ومنطقة بلاد ما بين النهرين المنخفضة، كما يوجد في البقع الرملية والحجر الجيري وسفوح التلال المتكتلة والسهول الطينية الصلبة. ينمو النبات على ارتفاع 150-10 م؛ موسم الإزهار والإثمار: فبراير – يونيو.

استُخدم مُستخلص البذور بالإغلاء لأغراضٍ طبيةٍ لعلاج الإمساك المزمن والزحار العصوي المزمن وككمادة. سُجِّل النبات كعلفٍ ربيعيٍّ في المناطق الحدودية بين العراق ونجد.

Prosopis farcta (Banks & Sol.) J.F.Macbr.

Family Fabaceae

الفصيلة البقولية

Syrian mesquite

kharroob خروب; *shok*, *shauk* شوك; *kharnūb* خرنوب; *estiri*, *astri* استري

A branched, straggling shrub to 3 m (10 ft); stems and branches with scattered thorns. Leaves with 3–5 pairs of pinnae and each pinna with 8–12 pairs of small leaflets. Flowers pale yellow, in spikes on about 2 cm- (0.8 in)-long peduncles. Pod short, about 5 cm (2 in) long and 2.5 cm (1 in) wide, swollen dorsally, indehiscent. Seeds hard, unwinged.

Prosopis farcta is native from North Africa eastwards to India. In Iraq, it is common in the lower forest zone, in foothills and on alluvial plains in the desert region, in mountain valleys, in open forest, in moist places in the lower hills, in depressions and along river valleys on the alluvial plains; alt. up to 1,500 m (4,900 ft); fl. Apr.–July.

The pods of *Prosopis farcta* have been used to treat swellings, as an anti-dysenteric and as an antispasmodic. According to Mati & de Boer (2011), it has been used as a purgative: 1 teaspoon of ground pods are eaten, and 1 teaspoon taken before food once a day for a month to treat colonitis.

شجيرة متفرعة منتشرة انتشارًا عشوائيًا ويبلغ طولها 3 أمتار. السيقان والفروع ذات الأشواك متناثرة. الأوراق بها 3–5 أزواج من الفلقات الريشية ولكل فلقة ريشية 8–12 زوجًا من الوريقات الصغيرة. الأزهار صفراء شاحبة، تتخذ شكل سنابل تنمو على سويقات طولها نحو 2 سم. الأجربة قصيرة، يبلغ طولها حوالي 5 سم وعرضها 2.5 سم، منتفخة ظهريًا، غير متفتحة. البذور صلبة، غير مجنحة.

الموطن الأصلي لنبات *Prosopis farcta* (الخروب) هو المنطقة الممتدة من شمال أفريقيا شرقًا إلى الهند. في العراق، ينتشر النبات في منطقة الغابات المنخفضة وسفوح التلال وعلى السهول الرسوبية في المنطقة الصحراوية وفي الوديان الجبلية والغابات المفتوحة وفي الأماكن الرطبة في التلال المنخفضة والمنخفضات وبامتداد وديان الأنهار في السهول الرسوبية. ينمو النبات على ارتفاع يبلغ 1500 م؛ موسم الإزهار:

استُخدمت أجربة نبات *Prosopis farcta* (الخروب) لعلاج التورُّمات، وكمُضادٍ للزحار والتشنج. وقد أورد كلٌ من ماتي ودي بوير أن النبات قد استُخدم كمُلينٍ حيث يتم تناول 1 ملعقة صغيرة من الأجربة المطحونة وملعقة صغيرة قبل الأكل مرة واحدة يوميًا لمدة شهر لعلاج التهاب القولون.

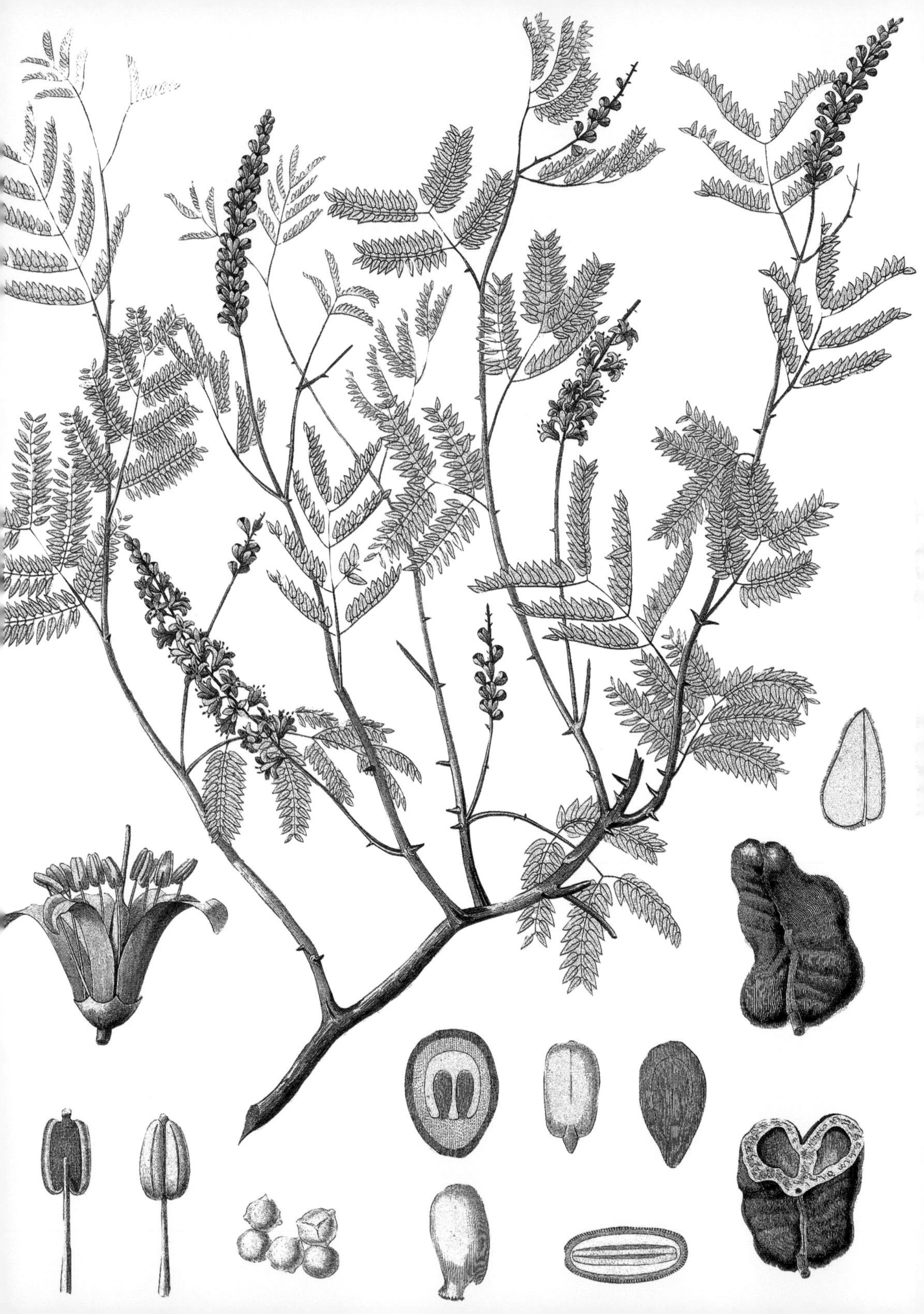

Punica granatum L.

Family Lythraceae

الفصيلة الخثرية

pomegranate

rummān رمان

A small tree or shrub, 1.5–5 m (5–16 ft) tall. Leaves opposite, oblong-lanceolate to obovate or elliptic. Sepals united, lobed above, reddish, somewhat succulent. Petals red, wrinkled in bud. Fruit reddish pink, pale red to scarlet, or brownish, with a thick rind, globose, 2–8 cm (0.8–3 in) in diameter, crowned with the lobes of the calyx, partitioned inside by thin, leathery yellow septa. Seeds red or pink, angular, with a juicy flesh.

Pomegranate is native from northeast Turkey east to northwest Pakistan; it is widely cultivated elsewhere. In Iraq, it is found in the mountains, in rocky places, and also in cultivation in the forest zone and lower hills; alt. 1,000–1,300 m (3,300–4,250 ft); fl. & fr. Aug.–Sep.

In Iraq, pomegranate is cultivated in gardens and orchards for its fruit, eaten fresh or as juice extracted as a cooling drink and tonic. A local variety with small black fruits that are inedible (*imharmal* أمحرمل) is used medicinally for the treatment of diarrhoea (rind), as a cooling drink (fruit juice), as an astringent and also as an anthelmintic, especially against tapeworms (bark).

The pomegranate is one of the oldest fruits known to humankind. The name NURMÛ (Akkadian) is recorded from ancient Mesopotamia. Excavations of sites from the Early Bronze Age (3500–2000 BCE) show pomegranate as one of the first cultivated fruits in southwest Asia (together with grape, olive and date palm). It has been used medicinally as a health tonic, its rind and bark as a vermifuge, and its juice as an antiseptic. Dried seeds of sour pomegranates have been used to help in dysentery, and the juice from seeds boiled and mixed with honey has been considered good for ulcers, sores and earache. The flowers have been used as an astringent and also as a red dye, and the rind for tanning leather. Pomegranate has always been a

شجرة صغيرة أو شجيرة طولها من 5 إلى 5 م. الأوراق متعاكسة، ومستطيلة-رمحية إلى بيضاوية مقلوبة أو إهليلجية. السبلات متحدة ومفصصة من الجهة العلوية، لونها ضارب إلى الحمرة، وعصارية إلى حد ما. البتلات حمراء اللون، مجعدة من عند البرعم. الثمرة وردية اللون مائلة إلى الحمرة، أو يتراوح لونها بين الأحمر الباهت وقرمزي، أو بنية اللون. الثمرة ذات قشرة سميكة، كروية الشكل، قطرها 8–2 سم، متوجة بفصوص من الكأس، ومُقسَّمة من الداخل بحواجز رقيقة جلدية صفراء اللون. البذور حمراء أو وردية اللون، زاوية، ذات لحم عصاري.

الموطن الأصلي لنبات الرمان هو المنطقة الممتدة من شمال شرق تركيا شرقًا إلى شمال غرب باكستان. يُزرع النبات على نطاق واسع في أماكن أخرى. في العراق، يوجد النبات في الجبال والأماكن الصخرية، كما يوجد في المزارع في منطقة الغابات والتلال المنخفضة. ينمو النبات على ارتفاع 1000–1300 م؛ موسم الإزهار والإثمار: أغسطس–سبتمبر.

في العراق، يُزرع الرمان في الحدائق والبساتين لثماره الطازجة أو العصير المُستخلص منه كمشروبٍ باردٍ ومُنشطٍ. يوجد صنف محلي له ثمار سوداء اللون وصغيرة الحجم غير صالحة للأكل (أمحرمل (imharmal) مُسجَّلة لاستخدامها طبيًا لعلاج الإسهال (القشرة)، وكمشروبٍ مُبرِّدٍ (عصير فواكه) وكمادةٍ قابضةٍ للمسام، وكمُضادٍ للديدان، لا سيّما الديدان الشريطية (لحاء النبات).

يُعد الرمان من أقدم الثمار التي عرفتها البشرية، وقد عُرف باسم NURMÛ (في اللغة الأكادية) في بلاد ما بين النهرين قديمًا. وتُبيِّن حفريات مواقع العصر البرونزي المُبكِّر (3500–2000 ق.م.) أن الرمان أحد ثمار الفاكهة الأولى المزروعة في جنوب غرب آسيا (وكذلك العنب والزيتون والنخيل). وقد استُخدم النبات طبيًا كمُنشطٍ صحيٍّ، بينما استُخدمت قشرته ولحاؤه كطاردٍ للديدان، واستُخدم عصيره كمُطهرٍ. استُخدمت البذور المجففة للرمان الحامض للمساعدة في علاج الزحار، وقد لوحظ أن عصير البذور المغلي والمخلوط بالعسل مفيدٌ في علاج التقرّحات وآلام الأذن.

fruit celebrated in many cultures as a symbol of fertility, prosperity, righteousness, eternal life and strength. It was esteemed by the Zoroastrians as a symbol of eternal life, and was a symbol of strength in Persian culture. It is believed to have been recommended by Prophet Muhammad as a fruit that purged the body and spirit of jealousy and hate. Pomegranate is one of the fruits mentioned in the Qur'an [Sura 6 (*An 'am* – Cattle) *Āyah* 99, 11; Sura 55 (*Rahman* – (God) Most Gracious) *Āyah* 68], referring to it as a blessing, as a symbol of paradise and as an example of avoiding excess and waste.

استُخدمت الأزهار كدواءٍ قابضٍ للمسام، كما استُخدمت الأزهار أيضًا كصبغةٍ حمراء اللّون والقشرة لدباغة الجلود. لطالما كان الرمان فاكهة مشهورة في العديد من الثقافات كرمزٍ للخصوبة والازدهار والصلاح والحياة الأبدية والقوة. احتفى الزرادشتيون بالثمرة؛ حيث اعتبروها رمزًا للحياة الأبدية، كما كانت رمزًا للقوة في الثقافة الفارسية. يُعتقد أن النبي محمد (صلّ الله عليه وسلم) قد أوصى بالرمان كفاكهةٍ تُطهر الجسد والروح من الغيرة والكراهية. ورد ذكر الرمان أيضًا في القرءان الكريم [السورة رقم 6 (الأنعام) الآيتان 99 و11؛ والسورة رقم 55 (الرحمن) الآية 68] في إشارة للنعيم ورمز للجنة وكمثالٍ على البعد عن الإسراف والسفه.

Rhus coriaria L.

Family Anacardiaceae

الفصيلة البطمية

sumac

summāq سماق

A small tree or shrub 1–5 m (3.3–16.4 ft) tall with white sap. Leaves with 9–17 leaflets, coarsely serrate at margins. Flowers unisexual, greenish-white, small, many in spreading terminal inflorescences; female flowers in separate shorter inflorescences. Fruit globose, 5–6 mm (0.2–0.24 in) in diameter, red, glandular-hairy.

Sumac is native to Macaronesia, and the Mediterranean Basin east to Afghanistan. In Iraq, it is common in the lower forest zone, in the mountains, in pine and oak forests and in rocky places on limestone; it is often planted as a hedge and in vineyards; alt. 550–1,300 m (1,800–4,300 ft); fl. & fr. July–Oct.

The fruit, acid to taste, has been used medicinally to stop bleeding, as a gargle for catarrh, and for the treatment of dysentery. An infusion of powdered or whole dried fruits, taken once or twice a day, has been recommended for the treatment of abdominal pain. Powdered dried fruits of sumac are widely used as a condiment sprinkled over grilled meat. The leaves produce a dye. The white sap of sumac is poisonous, however, and the plant should be used with caution.

شجرة صغيرة أو شجيرة يبلغ طولها 1–5 أمتار ذات نسغ أبيض اللون. الأوراق بها 9–17 وريقة، ولها حواف مسننة خشنة. الأزهار أحادية الجنس، بيضاء مائلة للخضرة، وصغيرة الحجم. الأزهار كثيرة منتشرة في نورات طرفية. الأزهار الأنثوية موجودة في نورات منفصلة أقصر طولًا. الثمرة كروية، يبلغ قطرها 5–6 مم، حمراء اللون، غدية، مزغبة.

الموطن الأصلي لنبات السماق هو ماكارونيسيا وحوض البحر الأبيض المتوسط شرقًا إلى أفغانستان. في العراق، ينتشر السماق في منطقة الغابات المنخفضة، وفي الجبال، وفي غابات الصنوبر والبلوط، وفي الأماكن الصخرية على الحجر الجيري. غالبًا ما يُزرع السماق كسياجٍ وفي مزارع الكروم. ينمو النبات على ارتفاع 550–1300 م؛ موسم الإزهار والإثمار: يوليو – أكتوبر.

استُخدمت الثمرة، حامضة المذاق، طبيًا لوقف النزيف، وكغرغرة لعلاج التهاب القناة التنفسية، ولعلاج الزحار. يُنصح بتناول منقوع مسحوق الثمار المُجففة أو باستخدام الثمار المُجففة الكاملة مرة أو مرتين يوميًا لعلاج آلام البطن. تُستخدم ثمار السماق المجففة المسحوقة على نطاق واسع كتابلٍ يُرش على اللحوم المشوية. تُنتِج الأوراق صبغة. النسغ الأبيض للسماق سام ويجب توخي الحذر عند استخدام النبات.

Ricinus communis L.

Family Euphorbiaceae

الفصيلة الفربيونية

castor oil plant

خروع *khirwa*; زيت الخروع

A large annual herb, with a hollow, single or branched stem, up to 5 m (16 ft). Leaves are commonly green, but purple forms also occur; the leaf-blade is 7–9-lobed with serrate or biserrate margins. Male and female flowers are reduced, and borne separately on the same plant; male flowers are yellowish-green, borne on the lower half of the inflorescence; female flowers are purplish, borne on the upper half. Fruit trilobed, smooth or sparingly to densely covered with bristle-tipped fleshy processes. Seeds glossy, greyish, silvery, or beige generally streaked and flecked with brown.

The castor oil plant is native to northeast tropical Africa; it has been introduced or cultivated in most warm tropical and temperate countries, and is often found as an escape, especially near villages. In Iraq, it grows on the irrigated alluvial plains in the desert region and is often found as an escape from cultivation.

The seeds yield the well-known castor oil, which has been used since ancient times in medicine as a purgative, an illuminant, a lubricant for mechanical objects and a preservative for leather. According to the Babylonian Talmudic school, in the story Jonah and the Whale, this was the plant species that provided shade to Jonah when he was cast onshore and was shaded by a plant.

In Iraq, castor oil has been recorded as a purgative and as an oil to strengthen hair. In the Arabian Peninsula, the seed oil has also been used as a purgative, and the oil massaged to relieve rheumatic pain. Leaves and roots have been used for treating bad breath, blisters and ulcers, toothache and inflamed eyes. It is a poisonous plant that should be used with caution.

عشبة حولية كبيرة الحجم، ذات ساق مجوفة، مفردة أو متفرعة، يبلغ طولها 5 م. عادة ما تكون الأوراق خضراء، ولكن تظهر أيضًا أزهار أرجوانية وأشكال أخرى. نصل الأوراق له 7 إلى 9 فصوص وحوافه مسننة أو مسننة مزدوجة. الأزهار الذكرية والأزهار الأنثوية مخفضة، وتُحمل بشكل منفصل على نفس النبات. الأزهار الذكرية خضراء مائلة للصفرة، ومحمولة على النصف السفلي من النورة، والأزهار الأنثوية أرجوانية اللون، ومحمولة على النصف العلوي. الثمرة ثلاثية الفصوص، ناعمة، ومغطاة على نحو معتدل إلى كثيف بحدبات لحمية ذات رؤوس خشنة. البذور لامعة، ورمادية، أو فضية، أو بنية فاتحة اللون وبوجه عام مخططة ومرقطة باللون البني.

الموطن الأصلي لنبات زيت الخروع هو شمال أفريقيا الاستوائية. وقد استُقدم الخروع وزُرع في معظم البلدان الاستوائية والمعتدلة الدافئة، وغالبًا ما يوجد كنبتة مزروعة بدأت تنمو بكثرة بسبب الشتلات الطوعية أو النمو غير الخاضع للرقابة، لا سيّما بالقرب من القرى. في العراق، يوجد النبات في السهول الرسوبية المروية في المنطقة الصحراوية، وغالبًا ما يُزرع ويوجد الحدائق.

تُنتِج البذور زيت الخروع المعروف الذي استُخدم منذ القدم في الطب كعامل تنقية ووسيلة إنارة ومادة مُزلقة للأغراض الميكانيكية وكمادةٍ حافظةٍ للجلد. ووفقًا لكتاب التلمود البابلي، في قصة يونس والحوت، كان هذا هو نوع النبات الذي استظل به يونان (يونس) عندما ألقاه الحوت على الشاطئ.

في العراق، سُجِّل زيت الخروع كمُلينٍ ولتقوية الشعر. وفي شبه الجزيرة العربية، يُستخدم زيت بذور الخروع كمُلينٍ وفي التدليك لتخفيف الآلام الروماتيزمية. وقد استُخدمت الأوراق والجذور في علاج رائحة الفم الكريهة والبثور والقروح وآلام الأسنان والتهاب العيون. يُعد الخروع نباتًا سامًا ويجب توخي الحذر عند استخدامه.

Senna alexandrina Mill.

Family Fabaceae

الفصيلة البقولية

Cassia senna L.; *C. alexandrina* (Mill.) Spreng.
Alexandrian senna, Arabian senna

sanā makki سنا مكي

A shrub 0.3–3 m (1–3.3 ft) tall. Leaves with 4–8(–12) pairs of lanceolate to narrowly elliptic or elliptic leaflets, 5–15 cm (2–6 in); stipules linear. Flowers yellow, in racemes. Stamens 10; filaments with 2 large, 5 medium and 3 small anthers. Pods 4–7 cm (1.6–2.8 in) long, flattened, dehiscent, transversely septate, papery. Seeds netted or wrinkly, with a small areole on each face.

Senna is native from the Sahara and Sahel regions of Africa eastwards to the Indian Subcontinent. In antiquity, *sanā* was exported from Egypt through the port of Alexandria, from where it was sent to Andalusia and elsewhere in Europe, hence the name Alexandrian Senna. The main medicinal use of senna was as a purgative (leaves and pods) but it was also used to treat epilepsy and chicken pox (Amar & Lev, 2017: p. 117). In Iraq, senna is an introduced plant, recorded as being cultivated in the desert region. There the seeds have been used to treat constipation and stomach cramps.

Another related species, *Senna italica* (*'ishriq* عشرق) found in Iraq (distributed in North and West Africa and throughout the Arabian Peninsula) is recorded to have medicinal properties similar to those of *S. alexandrina*. *Senna italica* is similar in appearance: a shrub up to 60 cm (24 in), with erect to ascending branches; leaves with 2–6 pairs of leaflets, yellow flowers in terminal and axillary racemes and a smaller pod, 3–5 cm (1.2–2 in) long, flattened, curved with a central crest.

شجيرة طولها 0.3–3 م. الأوراق ذات 4–8 (–12) أزواج من الوريقات الرمحية إلى الإهليلجية ضيقة أو إهليلجية، يبلغ طولها 5–15 سم؛ ذات أذينات خطية. الأزهار صفراء اللون، في عناقيد. 10 أسدية خيطية بها 2 من المآبر الكبيرة و5 من المآبر المتوسطة و3 من المآبر الصغيرة. طول الأجربة 4–7 سم، مفلطحة، ومتفتقة، وذات حواجز عرضية، وورقية. البذور ملمسها شبكي أو مجعدة، مع وجود نتوء صغير على كل وجه.

الموطن الأصلي للنبات هو مناطق الصحارى في أفريقيا وساحلها، شرقًا إلى شبه القارة الهندية. في العصور القديمة، كانت نبات السنا يُصدَّر من مصر عبر ميناء الإسكندرية، إلى الأندلس وأماكن أخرى في أوروبا، ومن هنا اكتسب اسمه «السنا الإسكندري» (Alexandrian Senna). استُخدم السنا طبيًا على نحو رئيسي كمُلينٍ (الأوراق والأجربة)، ولكنه كان يُستخدم أيضًا لعلاج الصرع والجديري الماء (آمار وليف 2017). في العراق، استُقدم السنا وسُجِّل كنباتٍ يُزرع في المنطقة الصحراوية، وقد استُخدمت البذور لعلاج الإمساك وتقلصات المعدة.

سُجِّل نوعٌ آخر ذو صلة موجود في العراق واسمه *S. italica* (عشرق ('ishriq) (مُوزَّع في شمال وغرب أفريقيا وفي جميع أنحاء شبه الجزيرة العربية) لما عُرف عنه من الخصائص الطبية المشابهة لخصائص نبات *S. alexandrina*. يتشابه *S. italica* في المظهر؛ فهو شجيرة يبلغ طولها 60 سم، ذات فروع منتصبة إلى صاعدة. الأوراق ذات 2–6 أزواج من الوريقات، وأزهار صفراء في الطرف وعناقيد إبطية وجراب أصغر، يبلغ طوله 3–5 سم، مفلطح، ومنحني مع قمة مركزية.

Teucrium chamaedrys L.

Family Lamiaceae

الفصيلة الشفوية

common germander, horsechire, wall germander

kamaderyos كمادريوس

A perennial herb, 10–60 cm (4–24 in), with a short woody base and decumbent to ascending stems. Leaves ovate, oblong to obovate, with crenate-dentate or lobulate margins. Flowers purple to pink-purple, in whorls of 4–8 flowers in dense or lax terminal racemes. Corolla tube short and hairy in the throat, split dorsally to near base of tube, appearing 1-lipped; lip 5-lobed with the middle lobe largest. Stamens with purple hairs on filaments and purple anthers. Nutlets blackish-red.

Common germander is native from northwest Africa and Europe east to Turkey, Iran and Iraq. In Iraq, it is found in open oak forests, on limestone cliffs and on rocky slopes; alt. 800–2,000 m (2,600–6,500 ft); fl. Sep.

Common germander has been used in Europe for numerous disorders including arthritis, gallbladder disorders, gout, poor appetite and weight loss; it has also been used as a tonic for fevers. It has been cited by Ibn al-Baitar (c. 1240 CE) as a diuretic. In Iraq, an infusion made from the leaves has been taken for abdominal complaints and as a diaphoretic, diuretic, stimulant, tonic, astringent, antiseptic and vermifuge.

عشبة مُعمرة، يبلغ طولها 10–60 سم، ذات قاعدة خشبية قصيرة وساق مستلقية إلى صاعدة. الأوراق بيضاوية، أو مستطيلة إلى بيضاوية مقلوبة، وذات حواف مسننة مستديرة-مثلمة أو مفصصة. الأزهار أرجوانية إلى أرجوانية وردية، في جدلات تتألف من 8–4 أزهار في عناقيد كثيفة أو متراخية طرفية. التويج أنبوبي قصير ومزغب من عند العنق، ومنقسم ظهريًا بالقرب من القاعدة مما يعطيه مظهرًا ذا شفة واحدة. الشفة خماسية الفصوص والفص الأوسط أكبر. الأسدية ذات شعر أرجواني على الخيوط والمتك أرجواني اللون. الثمار جوزية حمراء اللون داكنة.

الموطن الأصلي للنبات هو المنطقة الممتدة من شمال غرب أفريقيا وأوروبا شرقًا إلى تركيا وإيران والعراق. في العراق، يوجد النبات في غابات البلوط المفتوحة وعلى منحدرات الحجر الجيري وعلى المنحدرات الصخرية. ينمو النبات على ارتفاع 800–200 م؛ موسم الإزهار: سبتمبر.

استُخدم النبات في أوروبا لعلاج العديد من الأمراض؛ بما في ذلك التهاب المفاصل واضطرابات المرارة والنقرس وفقدان الشهية وخسارة الوزن. وقد استُخدم كمُلطفٍ للحمى، واستشهد به ابن البيطار (نحو عام 1240 م) كمُدرٍّ للبول. في العراق، استُخدم منقوع الأوراق لعلاج شكاوى البطن، وكمُعرقٍ، ومُدرٍّ للبول، ومُنبهٍ، ومُلطفٍ، وقابضٍ للمسام، ومُطهرٍ، وطاردٍ للديدان.

Trigonella foenum-graecum L.

Family Fabaceae

الفصيلة البقولية

fenugreek

helba, halba حلبة; *shimli* شملي (Kurd. باللغة الكردية)

An erect aromatic annual herb up to 30 cm (12 in) tall. Leaves trifoliate, leaflets obovate to oblong, dentate at the apex. Flowers white or pale yellowish-grey, solitary, or paired in the axils of leaves. Pods erect, 6–10 cm (2.4–4 in) long, straight or curved, somewhat flattened, tapering to a beak, 10–20-seeded.

Fenugreek is native to Iraq, Iran, Afghanistan and Pakistan, and widely introduced and cultivated elsewhere. In Iraq it is found in the lower forest zone and on irrigated alluvial plains in the desert region; it is generally weedy in fields and found as an escape; alt. up to 900 m (2,950 ft); fl. & fr. Mar.–May.

Fenugreek is among the oldest known and cultivated medicinal plants. Its name ŠAMBALILTU is mentioned in ancient Assyrian medical texts and it has been regarded as a medicinal plant in Ancient Egypt, Ancient Greece, Rome and India. Dioscorides (II–124) cites Ancient Egyptians calling fenugreek *itasin* and recommending the plant to induce childbirth. Fenugreek seeds have been found in the tomb of Tutankhamun, the ancient Egyptian Pharaoh king (reigned 1333–1323 BCE). Fenugreek has been cited by Ibn al-Baitar (c. 1240 CE) as useful in improving bladder function.

In Iraq the entire plant has been used as a demulcent and emollient, usually in combination with other herbs. It has been used for the treatment of colic, bronchitis and cough, and as an emollient in poultices for sprains, boils and wounds. Mati & de Boer (2011) reported that powdered seeds in water are used for diabetes, to enhance appetite, as a sedative and to treat hyperglycemia. The leaves have been used as a vegetable and also as fodder. Fenugreek is commonly cultivated and used throughout the Arabian Peninsula for medicinal and culinary purposes. It is considered to give strength and 'clean the stomach' after childbirth. It is commonly used in soups and sauces in Yemen, where it is believed to enhance appetite and prevent constipation.

عشبة حولية عطرية منتصبة يبلغ طولها 30 سم. الأوراق ثلاثية الوريقات، والوريقات بيضاوية مقلوبة إلى مستطيلة، مسننة من عند القمة. الأزهار بيضاء أو رمادية مصفرة باهتة، منفردة، أو مقترنة من عند محاور الأوراق. الأجربة منتصبة، طولها 10–6 سم، مستقيمة أو منحنية، ومفلطحة نوعًا ما، مستدقة إلى معقوفة، وتحتوي على 20–10 بذرة.

الموطن الأصلي للحلبة هو العراق وإيران وأفغانستان وباكستان، وقد استُقدم النبات وزُرع على نطاق واسع في أماكن أخرى. في العراق، يوجد النبات في منطقة الغابات المنخفضة، وفي السهول الرسوبية المروية في المنطقة الصحراوية. وعادة ما ينمو كحشيشة في الحقول ويوجد كنبتة مزروعة بدأت تنمو بكثرة بسبب الشتلات الطوعية أو النمو غير الخاضع للرقابة. ينمو النبات على ارتفاع يبلغ 900 م؛ موسم الإزهار والإثمار: مارس - مايو.

تُعد الحلبة واحدةً من أقدم النباتات الطبية المعروفة والمزروعة. ورد اسمها (ŠAMBALILTU في النصوص الطبية الآشورية القديمة، وقد أُعتبرت نباتًا طبيًا في مصر القديمة واليونان القديمة وروما والهند. استشهد ديسقوريدوس (الجز الثاني، 124) بالنبات، مشيرًا إلى أن المصريين القدماء أطلقوا عليه اسم (*itasin*) وأوصوا به لتحفيز عملية الولادة. عُثر على بذور الحلبة في مقبرة توت عنخ آمون، الملك الفرعوني المصري القديم (نحو عام 125–12 ق.م.). واستشهد ابن البيطار (نحو عام 1240 م) بالحلبة لتحسين وظيفة المثانة.

في العراق، استُخدمت النبتة بأكملها كمرهمٍ ومُلطفٍ، وعادة ما تُخلط مع أعشاب أخرى. وقد استُخدمت النبتة في علاج المغص والتهاب الشعب الهوائية والسعال وكمُلطفٍ في كمادات علاج الالتواءات والدمامل والجروح. أفاد ماتي ودي بوير بأن خليط مسحوق البذور مع الماء يُستخدم لعلاج داء السكري وتحسين الشهية وكمُهدئ ولعلاج ارتفاع السكر في الدم. استُخدمت الأوراق كخضراوات وأيضًا كعلفٍ. تشيع زراعة الحلبة واستخدامها في جميع أنحاء شبه الجزيرة العربية للأغراض الطبية والطهوية. وقد أُعتبرت نبتة مفيدة في تقوية الجسم و»تنظيف المعدة« بعد الولادة. يشيع استخدام الحلبة في الحساء والصلصات في اليمن؛ حيث يعتقد أنها تُحسِّن الشهية وتمنع الإمساك.

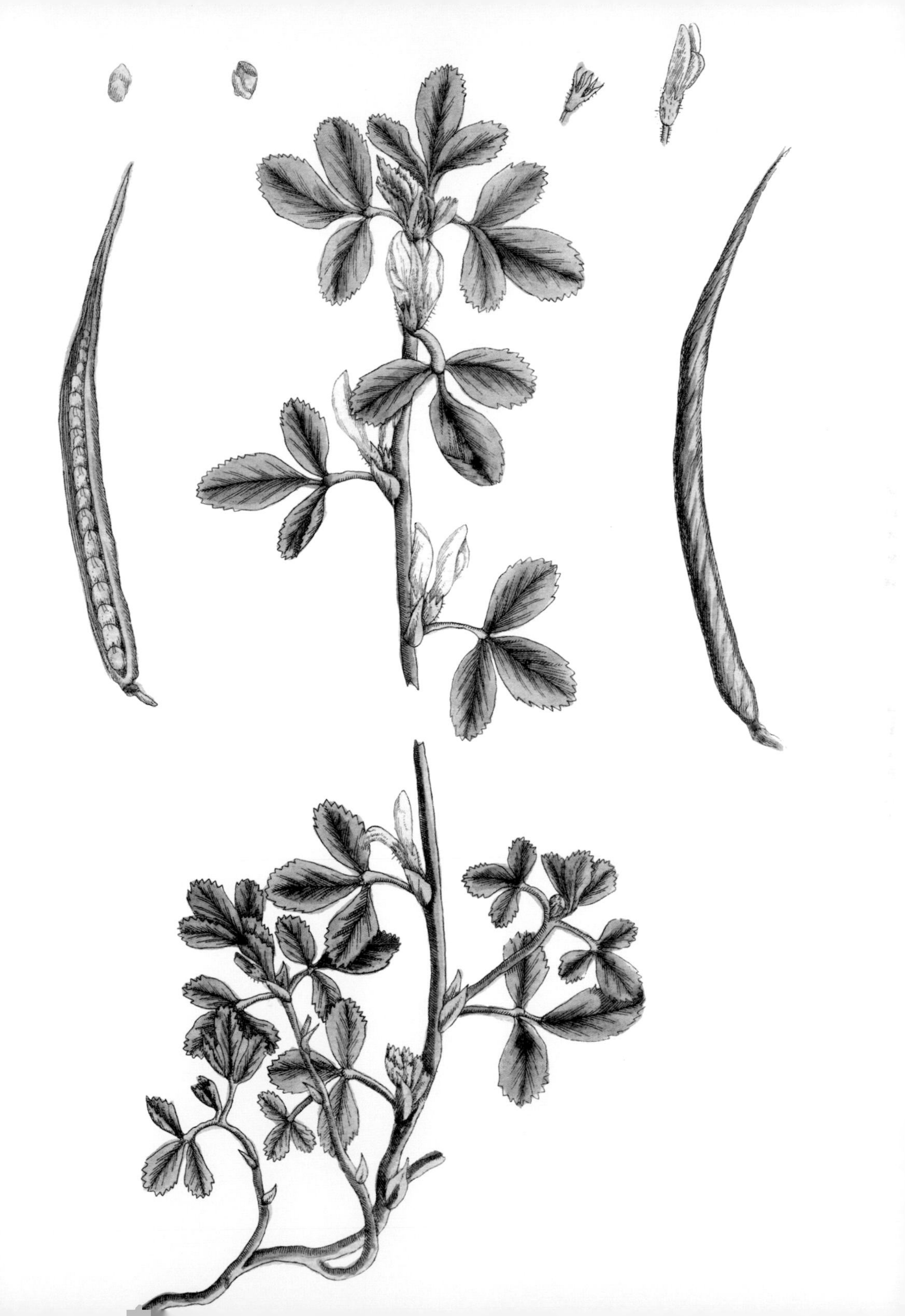

Urtica pilulifera L.

Family Urticaceae

الفصيلة القراصية

Roman nettle

قريص *qaris*; قريص روماني

An annual herb with stinging hairs, 30–75 cm (12–30 in) tall. Leaves broadly ovate, with coarsely dentate to serrate margins. Male and female flowers small, borne on separate inflorescences. Male flowers in dense clusters on branches of axillary inflorescences; female flowers in globose heads; inner segments of perianth of female flowers become accrescent (expanded) in fruit. Seeds (achenes) compressed, broadly ovoid, enclosed by the accrescent perianth.

Roman nettle is native to the eastern Mediterranean and northern Africa, to the Ukraine, Arabian Peninsula, Iran and Pakistan. In Iraq, it is found in the lower forest zone and lower hills, in shady locations on cliffs, on limestone in gorges and in waste places; alt. 150–1,000 m (500–3,300 ft); fl. & fr. Mar.–May.

Roman nettle has been cited by Ibn al-Baitar (c. 1240 CE) to dissolve kidney stones. In Iraq, an infusion of leaves used for the treatment of enlarged prostrate and alopecia is recorded by Mati & de Boer (2011).

عشبة حولية ذات شعر لاذع الملمس، يبلغ طولها 75–30 سم. الأوراق بيضاوية عريضة، والحواف مسننة إلى مسننة مستديرة خشنة. الأزهار الذكرية والأزهار الأنثوية صغيرة الحجم ومحمولة على نورات منفصلة. الأزهار الذكرية موجودة في مجموعات كثيفة على فروع النورات الإبطية، والأزهار الأنثوية موجودة في رؤوس كروية. الأجزاء الداخلية من غلاف الزهرة الأنثوية تصبح متصاعدة في الثمرة. البذور (الثمار) مضغوطة، بيضاوية عريضة، محاطة بغلاف الزهرة المتصاعد (الممتد).

الموطن الأصلي للنبات هو شرق البحر الأبيض المتوسط وشمال أفريقيا، إلى أوكرانيا وشبه الجزيرة العربية وإيران وباكستان. في العراق، يوجد النبات في منطقة الغابات والتلال المنخفضة، في مواقع مظللة على المنحدرات، على الحجر الجيري في الوديان وفي أماكن المهملة. ينمو النبات على ارتفاع 150–1000 م؛ موسم الإزهار والإثمار: مارس – مايو.

استشهد ابن البيطار (نحو عام 1240 م) بالنبات لتفتيت حصوات الكلى. في العراق، سُجِّل استخدام منقوع الأوراق في علاج تضخم البروستاتا والثعلبة بواسطة ماتي ودي بوير.

Herb. Horti Bot. Nat. Belg. (BR)
BR - S.P.
1 083 732
000010 83732
Herbier du Jardin botanique de Bruxelles
DU MORTIER

Visnaga daucoides Gaertn.

Family Apiaceae

الفصيلة الخيمية

Ammi visnaga (L.) Lam.
toothpick plant

خلة *khillah*; خيزران *khaizarān*; خلة بلدية

A stout, erect annual herb, 25 cm–1 m (10 in–3.3 ft) tall. Stems leafy, branching from near the base. Leaves divided into linear segments. Flowers small, white, grouped to form large umbels with many stout rays. Bracts at base of umbels divided into linear segments. Fruit ovoid to oblong, about 2 mm (0.008 in), with narrow longitudinal ribs.

The toothpick plant is native to the Mediterranean Basin, east to Iran and eastern Ethiopia. In Iraq it is common in the lower hills and grasslands, being found in moist places, by streams and ditches, on abandoned hillsides, in fields among stubble and in rice fields; alt. 50–600 m (160–2,000 ft), rarely to 1,500 m (4,900 ft) in the forest zone; fl. & fr. May–Aug.

In Iraq, a decoction of seeds of the toothpick plant has been used as a diuretic, as an antispasmodic and for the treatment of bladder stones. It has also been applied as a relaxant for muscles, for chest pains and to ease breathing in asthma. Dried stout rays of the umbels are commonly used as toothpicks.

A decoction of fresh seeds of a related species *Ammi majus* (*khella shaitāni* خلة شيتاني; *ghurair* غرير) has been used for vitiligo and as an emollient in Iraq.

عشب حولي سميك منتصب، يبلغ طوله 25 سم - 1 متر. الساق مورقة ومتفرعة من القاعدة القريبة. الأوراق مُقسَّمة إلى فصوص طولية. الأزهار صغيرة، بيضاء اللون، تتجمع لتشكل أزهار خيمية كبيرة ذات تفرعات قوية. القنابات عند قاعدة الأزهار الخيمية مُقسمة إلى فصوص طولية. الثمرة بيضاوية إلى مستطيلة الشكل، يبلغ طولها نحو 2 مم، ذات ضلوع طولية ضيقة.

الموطن الأصلي للخلة البلدية في حوض البحر المتوسط، وشرق إيران، وشرق إثيوبيا. يشيع وجود الخلة البلدية في العراق على التلال المنخفضة والأراضي العشبية والأماكن الرطبة إلى جوار الجداول والخنادق وسفوح التلال المهجورة، وفي الحقول بين القصب وحقول الأرز. ينمو العشب على ارتفاع 600–5 م. ويندر وجوده على ارتفاع 1500 م في منطقة الغابات؛ موسم الإزهار والإثمار: مايو - أغسطس.

في العراق، استُخدم مُستخلص بذور الخلة البلدية بالإغلاء كمُدرٍّ للبول ومضادٍ للتشنج وفي علاج حصوات المثانة، كما استُخدم النبات كباسطٍ للعضلات لعلاج آلام الصدر وتسهيل التنفس للحالات التي تعاني من الربو. يشيع استخدام الفروع السميكة الجافة للأزهار الخيمية كخلاتٍ للأسنان.

استُخدم مُستخلص البذور الطازجة لأحد الأنواع ذات الصلة، وهو الخلة الكبير (*Ammi majus*) (خلة شيتاني؛ غرير)، بالإغلاء كعلاجٍ للبهاق وكدهانٍ لطيفٍ للبشرة في العراق.

Withania somnifera (L.) Dunal

Family Solanaceae

الفصيلة الباذنجانية

Indian ginseng

;ورق الشفاء *waraq al shifa* ;جنسنج هندي
سم الفراخ *samm al ferakh* ;سيكران *saykarān*

Small shrub or woody-based herb to 2 m (6.6 ft) tall. Leaves greyish-green, ovate to obovate or oblong. Flowers small, greenish, in 3–6-flowered axillary clusters. Fruit a globose berry, red, shiny, surrounded by enlarged calyx lobes. The root is reported widely to have a strong, horse-like aroma.

Indian ginseng is widely distributed in the drier parts of tropical and subtropical regions of the Old World, from the Canary Islands, Mediterranean Basin and Africa, east to the Middle East, Sri Lanka, India and China. In Iraq, it is found in the lower hills and desert regions, in disturbed places, along roadsides, and at the edges of fields; alt. 250–500 m (820–1,640 ft); fl. & fr. Apr.–Aug.

Indian ginseng is much regarded for its stress-alleviating properties and has been widely used as a narcotic and sedative. The Arabic name *saykarān* is believed to be derived from Proto-Semitic *škr* (Arabic *skr*) 'to be intoxicated, inebriated', alluding to the sleep-inducing properties of the plant. It has a long history of use and was known to the Ancient Egyptians; fruiting branches have been found in the burial floral collar of Tutankhamun, the Egyptian Pharoah king (reigned 1333–1323 BCE). It is known as the 'Indian Ginseng' for its multiple medicinal uses and long history in traditional medicine in India. In India it is grown as a medicinal crop, with the whole plant or its different parts used in Ayurvedic and Unani systems of medicine; it is mentioned as an official drug in the Indian Pharmacopoeia.

شجيرة صغيرة أو عشب ذو قاعدة خشبية، يبلغ طوله 2 م. الأوراق خضراء مائلة إلى اللون الرمادي، وبيضاوية الشكل إلى بيضاوية مقلوبة أو مستطيلة. الأزهار صغيرة، مخضرة، في مجموعات إبطية مزهرة مكونة من ٣ إلى 6 أزهار. الثمار تشبه التوت وكروية الشكل، وحمراء اللون لامعة، محاطة بفصوص الكأس المتضخمة. من الشائع على نطاق واسع أن الجذر له رائحة قوية شبيهة برائحة الأفراس.

ينتشر الجنسنج الهندي على نطاق واسع في الأجزاء الجافة من المناطق الاستوائية وشبه الاستوائية في العالم القديم، من جزر الكناري وحوض البحر الأبيض المتوسط وأفريقيا، شرقًا إلى الشرق الأوسط وسريلانكا والهند والصين. في العراق، يوجد النبات في التلال المنخفضة والمناطق الصحراوية وفي التربة المضطربة وعلى جوانب الطرق وعند أطراف الحقول. ينمو النبات على ارتفاع 250–500 م؛ موسم الإزهار والإثمار: إبريل–أغسطس.

يُشار كثيرًا إلى الجنسنج الهندي لما يتمتع به من خصائص تُخفِّف من التوتر وقد شاع استخدامه على نطاق واسع كمُخدرٍ ومُسكِّن. يُعتقد أن الاسم العربي سيكران مشتق من اللغة السامية الأولى (*škr*) (بالعربية سكر) «تعني أن يكون أحدهم مخمورًا أو ثملًا»، في إشارة إلى خصائص النبات المهدئة التي تساعد على النوم. وقد حظي استخدام هذا النبات بتاريخ طويل؛ حيث عُرف لدى القدماء المصريين؛ وعُثر على الفروع المثمرة في القلادة الزهرية للدفن والخاصة بتوت عنخ آمون، أحد الملوك الفراعنة المصريين (حكم في الفترة 1333–1323 ق.م.). يُعرف النبات باسم «الجنسنج الهندي» نظرًا لاستخداماته الطبية المتعددة واستخداماته الطويلة في الطب التقليدي في الهند. وفي الهند، يُزرع النبات كمحصولٍ طبيٍّ، مع النبات بأكمله أو أجزائه المختلفة المستخدمة في أنظمة الطب الأيورفيدي واليوناني. ذُكر النبات كدواءٍ رسميٍّ في دستور الصيدلة الهندي.

In Iraq, the leaves have been used as a poultice on tumours; the roots have been used for reducing inflammation, psoriasis, bronchitis, and on ulcers; the fruit has been used as a diuretic and as a mild sedative. In the Arabian Peninsula, pounded leaves have been used as a poultice on burns and sunburnt skin; mixed with garlic, it is applied on stings and bites. In Africa, root extracts have been used to treat intestinal worms, stomach disorders, coughs, fevers and as an emetic, a diuretic and a tonic.

في العراق، استُخدمت الأوراق كمادةٍ للأورام، واستُخدمت الجذور لتقليل الالتهاب والصدفية والتهاب الشعب الهوائية والقرحة. استُخدمت الثمار كمُدرٍّ للبول وكمُسكّنٍ خفيف. وفي شبه الجزيرة العربية، استُخدمت الأوراق المطحونة كمادةٍ للحروق والبشرة المتعرضة لحروق الشمس، كما تُخلط الأوراق المطحونة مع الثوم ويوضع الخليط على اللسعات واللدغات. في أفريقيا، يُستخدم مُستخلص الجذور لعلاج الديدان المعوية، واضطرابات المعدة، والسعال، والحمى، كما يستُخدم كمُقَيِّئٍ ومُدرٍّ للبول ومُنشطٍ.

Zingiber officinale Roscoe

Family Zingiberaceae

الفصيلة الزنجبيلية

ginger

zanjābīl زنجبيل; *irq har* عرق حار

A rhizomatous herb with a thick, branched, fleshy rhizome (underground stem), strongly aromatic; rhizome yellowish inside. Shoots (pseudostems) arise from the rhizome, formed of bases of leaves, 50–100 cm (20–39 in) tall. Leaves lanceolate or linear-lanceolate, 15–30 cm (6–12 in). Inflorescence of spikes made from a series of greenish to yellowish leaf-like bracts. Flowers pale yellow with a purplish lip spotted with yellowish dots and streaks.

The Arabic name *zanjābīl* is based on an Indian vernacular (*sangavira*) corresponding to *siṅgivera* (Pali) and *çṛṅgavera* (Sanskrit). It is the name for ginger used in many parts of the Old World, including China where it is also referred to as the dried ginger of India, indicating where it originally came from.

Ginger is native to the maritime regions of Southeast Asia, the eastern Himalaya, Assam, India and south-central China. It is cultivated in most tropical countries, and in Iraq in the lower forest zone and alluvial plains. Ginger, along with other food plants and animals, spread throughout the Indo-Pacific some 5,000 years ago, carried by the Austronesian peoples as they migrated to the Indo-Pacific, reaching as far as Hawaii. Ginger was also one of the first spices exported from Asia, arriving in Europe with the spice trade, and was used by ancient Greeks and Romans. By the first century, ginger was known throughout the Mediterranean Basin, where it was used as a spice, as a medicine for bronchitis, as a carminative, as a stimulant and as an anti-emetic.

Ginger has been used as a medicine in Chinese, Ayurvedic and Greco-Arab medicine since antiquity. It is mentioned by Dioscorides (II–190) as a green herb used for many purposes; the roots (rhizomes) are warming and digestive, soften the intestines gently, and are good for the stomach. Maimonides describes it as the 'ginger of Syria' (*az-zanjābīl ash shāmī*), and as a tonic and stimulant.

عشب ذو جذامير سميكة متفرعة لحيمة (الساق تحت سطح الأرض) وذات رائحة قوية. السطح الداخلي للجذامير مصفر. البراعم (الجذوع الكاذبة) تنمو من الجذامير وتتكوّن من قواعد من الأوراق يبلغ طولها 100–50 سم. الأوراق رمحية أو رمحية خطية يبلغ طولها ١٥–٣٠ سم. النورات مُكوَّنة من سنابل عبارة عن سلسلة من القنابات شبيهة بالأوراق مائلة للخضرة إلى الصفرة. الأزهار صفراء باهتة ذات شفة أرجوانية مرقطة بنقاط وشرائط صفراء اللون.

الاسم العربي (زنجبيل) مُستوحى من العامية الهندية (sangavira) الذي يقابله (*siṅgivera*) (في لغة بالي)، و(*çṛṅgavera*) (في اللغة السنسكريتية)، وهو اسم الزنجبيل المُستخدم في أنحاء كثيرة من العالم القديم، بما في ذلك الصين؛ حيث يُشار إليه أيضًا باسم الزنجبيل المُجفف في الهند، في إشارة إلى مصدره الأصلي.

الموطن الأصلي للزنجبيل هو المناطق البحرية لجنوب شرق آسيا وشرق الهيمالايا وآسام والهند وجنوب وسط الصين. يُزرع النبات في معظم البلدان الاستوائية، وفي العراق، وفي منطقة الغابات المنخفضة والسهول الرسوبية. وقد انتشر الزنجبيل، وغيره من النباتات الغذائية والحيوانات، في جميع أنحاء منطقة المحيطين الهندي والهادئ منذ حوالي 5000 عام، حيث نُقل مع الشعوب الأسترونيزية أثناء هجرتهم إلى منطقة المحيطين الهندي والهادئ وصولًا إلى هاواي. كان الزنجبيل أيضًا أحد أوائل التوابل المُصدَّرة من آسيا، وقد وصل إلى أوروبا ضمن حركة تجارة التوابل، واستخدمه الإغريق والرومان القدماء. ومع حلول القرن الأول، كان الزنجبيل معروفًا في جميع أنحاء حوض البحر الأبيض المتوسط، ويستخدم كتابلٍ ودواءٍ لالتهاب الشعب الهوائية، وكطاردٍ للريح ومُنبهٍ ومُضادٍ للقيء.

استُخدم الزنجبيل كدواءٍ في الطب الصيني والأيورفيدي واليوناني العربي منذ القدم. وقد ذكره ديسقوريدوس (الكتاب 2–190) واصفًا إياه بأنه عشب أخضر متعدد الأغراض؛ (الجذامير) مُدفِّئة ومُهضِّمة، وتُستخدم لتليين الأمعاء بلطفٍ، وهي مفيدة للمعدة. يصف موسى بن ميمون الزنجبيل بلفظ «ginger of Syria» (الزنجبيل الشامي) مشيرًا إلى استخدامه كمُنشطٍ ومُنبهٍ.

In Iraq, the uses recorded by Mati & de Boer (2011) include keeping the body warm, helping with weight loss, treating flatulence, abdominal pain, coughs and pneumonia, use as a restorative, treating rheumatism, and improving blood circulation and sweating; mixed with *Lawsonia inermis*, it has been used as an antihypertensive agent, and mixed with sesame oil, it has been applied as a poultice on the body.

في العراق، سجّل ماتي ودي بوير استخدامات النبات والتي شملت الحفاظ على دفء الجسم، والمساعدة في إنقاص الوزن، وعلاج انتفاخ البطن، وآلام البطن، والسعال، والالتهاب الرئوي، وكمُقوٍّ في حالات الروماتيزم، وتحسين الدورة الدموية، والتُعرّق. يُمزج الزنجبيل مع الحناء (*Lawsonia inermis*) ويُستخدم كعاملٍ مُضادٍ لفرط الحساسية، كما يُخلط بزيت السمسم ويُستخدم الخليط ككمادةٍ للجسم.

FLORA OF IRAQ

Bryonia multiflora

Boiss. & Heldr.

Det. Date

Professor C. Haussknecht. Iter orientale 186

dieb.

IRAQ (MSU)

ROYAL BOTANIC GARDENS KEW

K000742792

Bryonia multiflora

BIBLIOGRAPHY فهرس

Abdullah Yusuf Ali, 1999 [1934–8]. *The Holy Qur'an: Text, Translation and Commentary.* [English].

Abd'ur Rehman as Sayuti, 1999. *Medicine of the Prophet.* Translated and edited by Ahmad Thompson. Ta Ha Publishers, London.

Abu-Irmaileh, B.E. & Afifi, F.U., 2003. Herbal Medicine in Jordan with Special Emphasis on Commonly Used Herbs. *Journal of Ethnopharmacology* 89: 193–7.

Ahmad, S.A. & Askari, A.A., 2015. Ethnobotany of the Hawraman Region of Kurdistan Iraq. *Harvard Papers in Botany* 20(1): 85–9.

Ahmed, H.M., 2016. Ethnopharmacobotanical Study on the Medicinal Plants Used by Herbalists in Sulaymaniyah Province, Kurdistan, Iraq. *Journal of Ethnobiology & Ethnomedicine* 12(1): 1–17.

Al-Baytār, Ibn, 1291. Al-Ŷāmi' li-mufradāt li-adwiya wa-l-agḏiya, 4 t., 2 vols. *Būlāq: Dār al-Madīna.*

Albayaty, N., 2011. The Most Medicinal Plants Used in Iraq: Traditional Knowledge. *Advances in Environmental Biology* 5(2): 401–6.

Al-Douri, N., 2014. Some Important Medicinal Plants in Iraq. *ASJ International Journal of Advances in Herbal and Alternative Medicine (IJAHAM)* 2(1): 10–20.

Al-Douri, N.A., 2000. A Survey of Medicinal Plants and Their Traditional Uses in Iraq. *Pharmaceutical Biology* 38(1): 74–9.

Al Ishbīlī, Abu Ḥayr, 2007. *Kitābu 'umdati tṭṭabīb fī ma 'rifati nnabāt likulli labīb*, vol. 2. Edición, notas y traducción castellana de J. Bustmente, F. Corriente y M. Tilmantine. CSIC, Madrid.

Al-Rawi, A., 1964. Wild Plants of Iraq with Their Distribution. *Technical Bulletin 14*, Directorate General of Agricultural Research and Project. Ministry of Agriculture, Government Press, Baghdad.

Al-Rawi, A. & Chakravarty, H.L., 1964. Medicinal Plants of Iraq. *Technical Bulletin 15*, Directorate General of Agricultural Research and Project. Ministry of Agriculture, Government Press, Baghdad.

Al-Snafi, A.E., 2018. Traditional Uses of Iraqi Medicinal Plants. *IOSR Journal of Pharmacy* 8(8): 32–96.

Amar, Z. & Lev, E., 2017. Arabian drugs in early medieval Mediterranean medicine. Edinburgh University Press, Edinburgh.

Amin, H.I., Ibrahim, M.F., Hussain, F.H., Sardar, A. Sh. & Vidari, G., 2016. Phytochemistry and Ethnopharmacology of Some Medicinal Plants Used in the Kurdistan Region of Iraq. *Nature Product Communications* 11(3): 291–6.

Boivin, N. & Fuller, D.Q., 2009. Shell Middens, Ships and Seeds: Exploring Coastal Subsistence, Maritime Trade and the Dispersal of Domesticates in and Around the Ancient Arabian Peninsula. *Journal of World Prehistory* 22(2): 113–80.

Boivin, N.L., Zeder, M.A., Fuller, D.Q., Crowther, A., Larson, G., Erlandson, J.M., Denham, T. & Petraglia, M.D., 2016. Ecological Consequences of Human Niche Construction: Examining Long-term Anthropogenic Shaping of Global Species Distributions. *Proceedings of the National Academy of Sciences United States of America* 11(2): 688–96.

Brewster, J.L., 2008. *Onions and Other Vegetable Alliums*. Crop Production Science in Horticulture, Vol. 15. CABI, Wallingford.

Chacravarty, H.L. & Jeffrey, 1980. Cucurbitaceae. In C.C. Townsend and E. Guest (eds), *Flora of Iraq* 4 (1): 191–208. Ministry of Agriculture & Agrarian Reform, Bagdad.

Chandra, R., Babu, D.K., Jadhav, V.T., Jaime, A. & Silva, T.D., 2010. Origin, History and Domestication of Pomegranate. *Fruit, Vegetable and Cereal Science and Biotechnology* 2: 1–6.

Chelalba, I., Benchikha, N., Begaa, S., Messaoudi, M., Debbeche, H., Rebiai, A. & Youssef, F.S., 2020. Phytochemical composition and biological activity of *Neurada procumbens* L. growing in southern Algeria. *Journal of Food Processing and Preservation* 44(10), e14774.

Christenhusz, M.J.M., Fay, M.F. & Chase, M.W., 2017. *Plants of the World: An Illustrated Encyclopaedia of Vascular Plants*. Royal Botanic Gardens, Kew.

Dafni, A., 2016. Myrtle (*Myrtus communis*) as a Ritual Plant in the Holy Land – A Comparative Study in Relation to Ancient Traditions. *Economic Botany* 70(3): 222–34.

Dafni, A., Lev, E., Beckmann, S. & Eichberger, C., 2006. Ritual Plants of Muslim Graveyards in Northern Israel. *Journal of Ethnobiology and Ethnomedicine* 2(1): 38.

De Vos, P., 2006. The Science of Spices: Empiricism and Economic Botany in the Early Spanish Empire. *Journal of World History* 17(4): 399–427.

Duke, J.A., 2008. *Duke's Handbook of Medicinal Plants of the Bible*. CRC Press, Boca Raton, FL.

Farooqi, M.I.H., 2003. *Plants of the Qur'an*, 6th edn. Sidrah Publishers, Lucknow.
Frankopan, P., 2015. *The Silk Roads. A New History of the World*. Bloomsbury, London.
Frey, W., Probst, W. & Kürschner, H. (eds), 1986. A synopsis of the vegetation in Iran. In: Contributions to the Vegetation of Southwest Asia pp. 9–43. L. Reichert, Wiesbaden, Germany.
Ghazanfar, S.A., 1994. *Handbook of Arabian Medicinal Plants*. CRC Press, Boca Raton, FL.
Ghazanfar, S.A., 2012. Medicinal Plants of the Middle East. In R. J. Singh (ed.), *Genetic Resources, Chromosome Engineering and Crop Improvement*, pp. 162–80. CRC Press, Boca Raton, FL.
Ghazanfar, S.A. & McDaniel, T., 2016. Floras of the Middle East: A Quantitative Analysis and Biogeography of the Flora of Iraq. *Edinburgh Journal of Botany* 73(1), 1–24.
Ghazanfar, S.A. & Edmondson, J.R. (eds), 2001–16. *Flora of Iraq*, vol. 5(1)–5(2). Royal Botanic Gardens, Kew.
Ghazanfar, S.A., Edmondson, J.R. & Hind, J.N. (eds), 2019. *Flora of Iraq*, vol. 6. Royal Botanic Gardens, Kew.
Guest, E,. 1933. Notes on plants and plants products, with their colloquial names in Iraq. *Department Agriculture Iraq Bulletin* 27: 1–111.
Guest, E. & Al-Rawi, A., 1966. *Flora of Iraq*, vol. 1. The Ministry of Agriculture, Baghdad.
Hepper, F.N., 1992. *Illustrated Encyclopedia of Bible Plants*. Inter-Varsity Press, London.
Hort, A. (trans.), 1916. *Theophrastus: Enquiry into Plants*. Loeb Classical Library 79. Harvard University Press, Cambridge, MA.
Ibn Qayyim al Jawziyya, 1998. *Medicine of the Prophet*. Translated and edited by Penelope Johnston. Islamic Texts Society, Cambridge.
Ibn Sina (Avicenna), 1025. *The Canon of Medicine* (Ar. القانون في الطب *al-Qānūn fī al-Ṭibb*).
Janick, J., Paris, H.S. & Parrish, D.C., 2007. The Cucurbits of Mediterranean Antiquity: Identification of Taxa from Ancient Images and Descriptions. *Annals of Botany*, 100: 1441–57.
Kawarty, A.M.A., Beh Çet, L. & ÇakilcioĞlu, U., 2020. An Ethnobotanical Survey of Medicinal Plants in Ballakayati (Erbil, North Iraq). *Turkish Journal of Botany* 44: 345–57.
Levey, M., 1966. *The Medical Formulary or Aqrābādhīn of al-Kindī*. University of Wisconsin Press, Madison, WI.
Levin, G.M., 2006. *Pomegranate*. Third Millennium Publishing.
Manniche, L., 1989. *An Ancient Egyptian Herbal*. British Museum Press, London.

Mathew. *Flora of Iraq* (8:1985, p. 2121)

Mati, E. & de Boer, H., 2011. Ethnobotany and Trade of Medicinal Plants in the Qaysari Market, Kurdish Autonomous Region, Iraq. *Journal of Ethnopharmacology* 133(2): 490–510.

Mildrexler, D.J. , Zhao, M. & Running, S.W., 2005. Where are the hottest spots on earth? *EOS* 87 (43): 467

Musselman, L.J., 2012. *A Dictionary of Bible Plants*. Cambridge University Press, Cambridge.

Naqishbandi, A., 2014. Plants used in Iraqi traditional medicine in Erbil-Kurdistan region. *Zanco Journal of Medical Sciences* 18 (3), 811–5.

Newberry, P.E., 1889. On Some Funeral Wreaths of the Græco-Roman Period, Discovered in the Cemetery of Hawara. *The Archaeological Journal* 46: 427–32.

Olson, D.M. , Dinerstein, E., Wikramanayake, E.D., Burgess, N.D., Powell, G.V.N. & Underwood, E.C. *et al.*, 2001. Terrestrial ecoregions of the world: a new map of life on Earth. *Bioscience* 51(11): 933–8.

Osbaldeston, T.A. (trans.), 2000. *Dioscorides De Materia Medica*. Ibidis Press, Johannesburg.

Postgate, J.N., 1987. Notes on Fruits in the Cuneiform Sources. *Bulletin on Sumerian Agriculture* 3: 115–44.

Ravindran, P.N. & Babu, K.N., 2016. *Ginger: The Genus Zingiber*. CRC Press, Boca Raton, FL.

Rivera, D., Matilla, G., Obón, C. & Alcaraz, F., [2011] 2012. *Plants and Humans in the Near East and the Caucasus. Ancient and Traditional Uses of Plants as Food and Medicine. An Ethnobotanical Diachronic Review*. Volume 2. Ministerio de Ciencia e Innovacion of Spain.

Robinson, B., 1985. Jonah's Qiqayon Plant. *Zeitschrift für die Alttestamentliche Wissenschaft* 97: 90.

Rosner, F., 1995. *Maimonides, Medical Writings, The Glossary of Drug Names*. Translated and annotated from Max Meyerhof's French edition. The Maimonides Research Institute, Haifa, Israel.

Saad, B. & Said, O., 2011. *Greco-Arab and Islamic Herbal Medicine: Traditional System, Ethics, Safety, Efficacy, and Regulatory Issues*. John Wiley & Sons, Inc.

Saḥīḥ al-Bukhārī, c. 86. *al-Jaami' al-Sahih al-Musnad al-Mukhtasar min Umuri Rasoolillahi wa sunanihi wa Ayyaamihi* (*The Abridged Collection of Authentic Hadith with Connected Chains Regarding Matters Pertaining to the Prophet, His Practices and His*

Times). Muhammad-Bin-Isma`il Al-Bukhari, 2019. *Encyclopedia of Sahih Al-Bukhari.* Vol. 1, Bk 12, No. 812. Arabic Virtual Translation Center.

Sahih Muslim (b. 817/18, d. 87/75) is a collection of hadith compiled by Imam Muslim ibn al-Hajjaj al-Naysaburi. His collection is considered to be one of the most authentic collections of the Sunnah of the Prophet, and along with Sahih al-Bukhari forms the 'Sahihain', or the 'Two Sahihs'. It contains roughly 7,500 hadith (with repetitions) in 57 books. https://sunnah.com/muslim. Translated by Abdul Hamid Siddiqui.

Schweinfurth, G., 1888. Further Discoveries in the Flora of Ancient Egypt 1. *Nature* 29: 12–5.

Schweinfurth, G.A., 1886. Reise in das Depressionsgebiet im Umkreise des Fajum im Januar, 1886. *Zeitschrift der Gesellschaft für Erdkunde zu Berlin* 21: 96–149.

Townsend, C. & Guest, R., 1974. *Flora of Iraq*, vol. 4. The Ministry of Agriculture and Agrarian Reform, Baghdad.

Townsend, C., Guest, R. & Al-Rawi, A. (eds), 1966–8. *Flora of Iraq*, vols 2 & 9. The Ministry of Agriculture, Baghdad.

Townsend, C., Guest, R. & Omar, S. (eds), 1980–5. *Flora of Iraq*, vols (1)–(2) & 8. The Ministry of Agriculture and Agrarian Reform, Baghdad.

Youssef, S. 2020. Endemic Plant Species of Iraq: From Floristic Diversity to Critical Analysis Review. *Journal of University of Duhok (Agric. and Vet. Sciences)* 23(2): 90–105.

Zohary, D. & Spiegel-Roy, P., 1975. Beginnings of Fruit Growing in the Old World. *Science* 187(17): 319–27.

Zohary, D., Hopf, M. & Weiss, E., 2012. *Domestication of Plants in the Old World: The Origin and Spread of Domesticated Plants in Southwest Asia, Europe, and the Mediterranean Basin*. 4th edition. Oxford University Press, Oxford.

Zorc, R.D.P., 1994. Austronesian Culture History through Reconstructed Vocabulary (An overview). In A.K. Pawley & M.D. Ross (eds), *Austronesian Terminologies: Continuity and Change*, pp. 541–94. Research School of Pacific and Asian Studies, Canberra.

Visnaga daucoides

INDEX

OF LATIN, ENGLISH COMMON AND VERNACULAR NAMES

Latin names in **bold**, English common names in regular font, synonyms in SMALL CAPITALS and vernacular names in *italics*.

فهرس لأسماء النباتات

IMAGE SOURCES مصادر الصور

© The Board of Trustees of the Royal Botanic Gardens, Kew

Anethum graveolens
Glycyrrhiza glabra
Haloxylon salicornicum
Matricaria chamomilla
Nigella sativa
Pistacia khinjuk
Plantago ovata
Prunus armeniaca
Quercus libani subsp. *branti*
Ricinus communis
Senna alexandrina
Teucrium chamaedrys
Withania somnifera
Zingiber officinale

© Oxford University Herbaria

Agrimonia eupatoria
Alhagi maurorum
Althaea officinalis
Anastatica hierochuntica
Anchusa azurea
Apium graveolens
Arctium lappa
Artemesia campestris
Asparagus officinalis
Astragalus tribuloides
Bacopa monnieri
Bellis perennis
Bidens tripartita
Brassica nigra
Bryonia multiflora
Calotropis procera
Capparis spinosa
Chrozophora tinctoria
Citrullus colocynthis
Clinopodium graveolens
Crocus sativus
Cyperus rotundus
Glaucium corniculatum
Inula helenium
Lysimachia arvensis
Marrubium vulgare
Myrtus communis
Oxalis corniculata
Papaver rhoeas
Peganum harmala
Phyla nodiflora
Prosopis farcta
Punica granatum
Rhus coriaria
Trogonella foenum-graecum
Visnaga daucoides

© C. J. Thorogood

Cistanche tubulosa
Eminium spiculatum
Glossostemon bruguieri

© Linnean Society of London

Ammannia baccifera

© Meise Botanic Garden

Urtica pilulifera

الله اكبر